Education and
the Environment

Education and the Environment

Creating Standards-Based Programs in Schools and Districts

GERALD A. LIEBERMAN

Harvard Education Press
Cambridge, Massachusetts

Library of Congress Control Number 2013944858

Paperback ISBN 978-1-61250-629-6
Library Edition ISBN 978-1-61250-630-2

Published by Harvard Education Press,
an imprint of the Harvard Education Publishing Group

Harvard Education Press
8 Story Street
Cambridge, MA 02138

Cover Design: Deborah Hodgdon
Cover Photo: ralucahphotography.ro/Flickr/Getty Images

The typefaces used in this book are Minion, Myriad Pro, and Myriad Condensed

Contents

Foreword

For decades, Gerald Lieberman has been at the forefront of environment-based learning. The concept, which has acquired several names over the years, is essentially this: children and young people learn best when their time in the classroom is augmented by experiences in the wider community—including the natural world beyond the schoolhouse walls.

I first became aware of Gerald Lieberman's pioneering work in the early 2000s while researching *Last Child in the Woods*, my book about the widening divide between children—indeed, people of all ages—and the rest of nature. "Nature-deficit disorder," as I define it in the book, describes the human costs of alienation from nature suggested by an ever-growing body of research. Among the losses: diminished use of the senses, attention difficulties, physical and emotional illnesses, a rising rate of myopia, child and adult obesity, and other maladies. Most of this research remains correlative, not causal, but it points in a definite direction: experiences in the natural world can have a profoundly positive impact on a child's ability to learn, to be healthy, and to be happy.

Lieberman, along with a handful of other thinkers at the time, gave me hope when the outlook for a societal reconnection with nature was especially gloomy. As director of the State Education and Environment Roundtable (SEER), a national effort to improve learning by integrating environment-based education into kindergarten through high school curricula, he initially worked with administrators of departments of education and educators in 150 schools in sixteen states, helping to identify model environment-based programs and examining how the students fared on standardized tests. SEER's seminal report, *Closing the*

Achievement Gap: Using the Environment as an Integrating Context for Learning, issued in 1998, stated that "ecosystems surrounding schools and their communities vary as dramatically as the nation's landscape," and therefore, "the term 'environment' may mean different things at every school; it may be a river, a city park, or a garden carved out of an asphalt playground." SEER's findings were broad but stunning: environment-based education produced student gains in social studies, science, language arts, and math; improved standardized test scores and grade-point averages; and developed skills in problem solving, critical thinking, and decision making.

For example: in Florida, Taylor County High School teachers and students used the nearby Econfina River to team-teach math, science, language arts, biology, chemistry, and the economics of the surrounding county; in San Bernardino, California, students at Kimbark Elementary School studied botany and investigated microscopic organisms and aquatic insects in an on-campus pond and vegetable garden; in Glenwood Springs, Colorado, high school students planned and supervised the creation of an urban pocket park, and city planners asked them to help develop a pedestrian mall and park along the Colorado River; and at Huntingdon Area Middle School in Pennsylvania, students collected data at a stream near the school. A teacher there, Mike Simpson, used that information to teach fractions, percentages, and statistics, as well as how to interpret charts and graphs. "I don't have to worry about coming up with themes for application problems anymore. The students make their own."

David Sobel, director of teacher certification programs at Antioch University New England Graduate School, did an independent review of the SEER research and similar studies, including one by the National Environmental Education and Training Foundation. These studies reported findings similar to Lieberman's. Students in these programs typically outperformed their peers in traditional classrooms. If reading skills are "the Holy Grail of education reform," as Sobel put it, environment-based or place-based education should be considered "one of the knights in shining armor."

In the years following the SEER report, Lieberman's case for environment-based education has been bolstered by additional research from others. Yet societal change and, in some cases, education reform efforts have widened the gap between children and the natural world and, at the same time, devalued the importance of independent, imaginative play. As a result, children's abilities for self-control, their mental and physical health, and their sense of who they are in the world have been lessened. While the 2001 No Child Left Behind Act appears to have produced impressive statistical gains in some school districts, the implementation of the act has arguably undermined its effectiveness. Too often, educators and school boards, in what they interpreted as the spirit of the act, canceled or reduced recess, dropped field trips, increased testing, and further confined students and learning within school walls.

Today, a convincing argument can be made to retain some aspects of the act, but many educators—along with parents, business leaders, pediatricians, conservationists, and other stakeholders in our children's futures —now call for what can be called a No Child Left *Inside* movement.

As *Education and the Environment* goes to press, we see hopeful signs, including the emergence of major campaigns with broad-based support to connect children and the rest of us to nature. We see increased interest among pediatricians, some of whom are "prescribing" nature time. In education, we see a growing number of nature-based schools, particularly preschools. We also see more and more school gardens and schoolyard wildlife habitat areas, and an increased awareness among teachers that getting students outside can help them learn and focus and improve their mental health and social skills. (As I travel around the country, I hear many teachers tell this tale, often with almost the exact same wording: "The troublemaker in class becomes the leader when I take the class outside." Not just a better-behaved student, the *leader*.) In his new book, Lieberman cautions that environment-based education is "not a magic bullet that will solve all of the world's problems." But surely we see increasing evidence that children's nature deficit, and our own, can drain life and spirit from the human experience and reduce our children's sense of wonder and their will to learn. Not all educators will agree with this premise, or with Lieberman's approach. Pedagogy evolves. Yet, as Howard Frumkin, dean of the University of Washington School of Public Health says when surveying related evidence in his field, "Yes, we need more research. But we know enough to act."

Much is at stake. Worldwide, as of 2008, there are now more people living in cities than in the countryside. This transformation is a huge and largely unremarked milestone in human history. The continued urbanization of our daily lives will bring one of two outcomes: either the human connection to nature will continue to fade, or we'll create a new kind of civilization, one in which schools, workplaces, homes, and neighborhoods connect people to nature in their daily lives. Cities within that civilization could, indeed, become an incubator of biodiversity and human health. Shaping that future will require a broader definition of green jobs to include careers that focus not only on conservation and energy efficiency, but also on connecting people to nature for better productivity, creativity, health, and happiness. Ideally, children will attain a deep understanding of what it means to be human beings as part of nature.

In September 2012, the World Conservation Congress of the International Union for the Conservation of Nature (IUCN), citing "adverse consequences for both healthy child development ('nature-deficit disorder') as well as responsible stewardship for nature and the environment in the future" with the growing separation of people, and especially children, from nature, passed a resolution titled "The Child's Right to Connect with Nature and to a Healthy Environment." This resolution urges that reversing this trend become a priority around the globe. To many of us, this acknowledgment is an indicator of good things to come. As I wrote in *Last Child in the Woods*, school should be more than a polite form of incarceration; it should be a portal to a wider world. Gerald Lieberman's ongoing work underscores the right of a whole child to feel and be fully alive.

—Richard Louv

Acknowledgments

The materials for this book were developed with the support and help of many individuals, including all of those involved in the development of the concepts underlying environment-based education, especially:

The Pew Charitable Trusts: President Rebecca Rimmel, Executive Vice President Joshua Reichert, and program officers Joy Horwitz and Ellen Wert;

Council of Chief State School Officers;

The wonderful staff of the State Education and Environment Roundtable (SEER), Grace Lieberman and Jayne Henn, as well as Linda Hoody, the codeveloper of the Environment as an Integrating Context for learning (EIC) Model;

Representatives of SEER's founding member agencies: William Andrews, California Department of Education; Don Hollums and Mary Gromko, Colorado Department of Education; Kathy Shea Abrams, Florida Office of Environmental Education; Robert Moore, Georgia Department of Education, and Deron Davis, Georgia Department of Natural Resources Environmental Protection Division; Kevin Collins, Idaho Department of Education, and Donny Roush, Idaho Environmental Education Association; Duane Toomson and Kathy McKee, Iowa Department of Education; Jane Eller, Kentucky Environmental Education Council; Gary Heath, Maryland State Department of Education; Gregg Swanzey and Polly Zajac, Massachusetts Southeastern Environmental Education Alliance, and Melissa Griffiths, Massachusetts Executive Office of

Environmental Affairs; Kathleen Lundgren, Minnesota Department of Education, and Pam Landers, Minnesota Office of Environmental Assistance; Bruce Marganoff and Arthur Mitchell, New Jersey Department of Education; John Hug, Ohio Department of Education; Patricia Vathis and Dean Steinhart, Pennsylvania Department of Education; Edward Falco, South Carolina Department of Education; Irene Pickhardt, Texas Education Agency; and Tony Angell, Washington Office of the Superintendent of Public Instruction.

With gratitude to all of my colleagues who worked on, guided, encouraged, and supported the development of the Education and the Environment Initiative (EEI) curriculum, especially Jennifer Rigby, and at the California state agencies, Andrea Lewis, Mindy Fox, Linda Moulton Patterson, Bonnie Bruce, Rebecca Williams, Christy Porter Humpert, and innumerable other staff members from the California Environmental Protection Agency's Office of Education and the Environment, who all played crucial roles throughout the development of the EEI. Equally supportive and important in the success of the EEI were State Superintendent of Education Jack O'Connell, and Tom Adams, Phil Lafontaine, Celeste Royer, and many others from the California Department of Education, State Board of Education, California Natural Resources Agency, San Luis Obispo County Office of Education, and California State University, Sacramento. A special thanks to the State of California's leadership, including Governors Gray Davis and Arnold Schwarzenegger, and Winston Hickox, Terry Tamminen, Alan Lloyd, and Linda Adams, secretaries of the California Environmental Protection Agency, all of whom stood behind the EEI and provided funding even during difficult economic times.

Additional thanks to all the teachers, administrators, and students who shared their experiences in implementing environment-based education at: Arabia Mountain High School (Georgia), Armuchee Elementary School (Georgia), Brundrett Middle School (Texas), Centreville Middle School (Maryland), Chariton Middle School (Iowa), Concrete Middle School (Washington State), Clay County High School (Kentucky), Desert Sands Unified School District (California), Dowling Urban Environmental School (Minnesota), Eisenhower Middle School (New Jersey), Glenwood Springs High School (Colorado), Huntington Area Middle School (Pennsylvania), J. R. Briggs Elementary School (Massachusetts), Jackson Elementary School (California), Larkspur Elementary School (Colorado), New Bedford Global Learning Charter School (Massachusetts), Pickens Middle School (South Carolina), Red Oak Middle School (Iowa), Rigby Junior High School (Idaho), Seven Generations Charter School (Pennsylvania), and Twin Lakes Elementary School (California).

I also want to thank Nancy Walser of the Harvard Education Press for her support and guidance through all stages of developing this book. Several other individuals also contributed to its development including: Colleen Gillard, developmental editor; Rebecca Voorhees, graphic designer; and Luke Woodman, artist.

My partner and wife, Grace Lieberman, and son, Michael Lieberman, both of whom were always there to support me through the many long days and nights.

Introduction

Throughout most of human history people have lived in direct contact with nature, growing their own food, raising or killing animals to eat, using trees and stone to build homes, and using water for irrigation, household purposes, and transportation. Since the beginning of time, and long before the existence of formal systems of education, the most important thing humans taught their children was how to survive by exploiting nature's resources. Not until about 150 years ago, when more people began to live in cities than in the country and education became more formally organized around the three R's, did the environment become something optional when it came to education.

Today, appreciating and understanding how to sustain the natural environment is so absent from the modern classroom that, despite the growing litany of concerns around climate change, population growth, and environmental destruction that will affect future generations, few states, districts, or schools have any kind of environmental education built into their curricula.

Over the past two decades, teaching about the environment has decreased as schools have had to focus increasingly on helping students master the skills and content represented in state standards and show their mastery each year on high-stakes standardized tests that cover only two subjects: math and English language arts. In large part, the decrease in learning about the environment resulted from the fact that states, with the exception of Pennsylvania and California, did not develop standards related to instruction about the environment or environmental issues. As a result of the shift in focus to standards-based instruction and associated budgetary changes, teaching and learning about the environment and environmental issues and how they affect humans and the natural world has decreased dramatically.[1]

Even though many educators express interest in integrating environmental content into formal education, they are challenged to find time in the school day that is not already scheduled to meet the increasingly complex demands of state content requirements. During the past twenty years, education across the United States has become focused almost exclusively on instruction and assessment driven by state and school district standards. These curricular requirements have filled the school day to such an extent that many teachers are hard-pressed to find the time to focus on material not explicitly required by these standards, not to mention time to include trips outside the classroom. The once common school field trips to parks, forests, ponds, beaches, or even farms and zoos, which for many urban dwellers represent their only contact with the natural world, have been eliminated from many school districts as a result of changes in instructional priorities and resulting budget cuts.[2]

One other important factor has contributed to the decrease in opportunities for students to learn about the environment in school. While they were once viewed as a nonpartisan concern for all Americans, over the past forty years there has been a shift in the political landscape that has caused many environmental issues to be viewed from a political perspective. This was much less of a concern in 1970, when there was widespread bipartisan support for the founding of the U.S. Environmental Protection Agency and the passing of the first National Environmental Education Act during President Nixon's administration. Unfortunately, one of the results of this societal shift is that many people and organizations now perceive environmental education as part of a political agenda rather than seeing it as "integrated into the whole system of formal education at all levels to provide the necessary knowledge, understanding, values, and skills needed . . . (as) a prerequisite for resolving serious environmental problems at the global level," as was defined at the 1977 UNESCO-sponsored Tbilisi Conference.[3]

AN URGENT MATTER

Both supporters and critics of environmentalism, however, should be alarmed that today there is little time in school for examining how the survival of the human species and our individual lives depends on the health and functioning of Earth's natural systems. Nowhere in the current definition of a comprehensive education is there value placed on understanding human dependence on natural systems and how human decisions and activities influence everything from the survival of plant and animal species to the Earth's atmosphere and climate.

Diminished school-based opportunities for students to learn about the environment are leaving future generations unprepared to face the critical challenges of our rapidly changing world. Ultimately, students' success—job prospects and ability to participate in a civil society and contribute solutions necessary for maintaining a healthy environment—depends on their ability to identify, analyze, and balance the multitude of factors that can affect the environment.

For students to succeed in life and be able to resolve the environmental challenges of the twenty-first century requires that they be able to function effectively in society's decision-making processes, and not just "know" about the environment per se. They need to learn how human social systems (economics, laws, culture, politics, etc.) function *and* how these systems interact with natural systems in the environment.

But standards-based instruction and learning about the environment need not be mutually exclusive. By connecting instruction to a school's local environment, teachers can engage students with authentic lessons that directly support their efforts to help students become proficient in academic content standards. By interweaving academic content through interdisciplinary instruction and connecting it to environmental themes, or "contexts," environmental study becomes not simply another add-on to academic studies, but an engaging, integrating medium for multiple subjects. This book will show how it's done.

TAPPING INTO DEMAND

The good news is that there is a willing audience seeking to learn more about the environment. Studies of parental interest, conducted by the National Environmental Education Foundation and Roper Reports in 1997 and 2001, indicate that 96 percent of parents believe that students should be taught about the environment in their classrooms.[4] The same studies also found high levels of support among the general public. This interest is also found at the college and university level, where enrollment in environmental science and engineering are at an all-time high, with schools of higher education expanding their program offerings in environmental science to interdisciplinary offerings in earth science, urban studies, architecture, law, and politics at both graduate and undergraduate levels.

Preparation for such college majors, however, just isn't happening at the K–12 level, which is unfortunate, since environmental education is relevant, dynamic, emotionally engaging, and provides rich opportunities for hands-on, authentic teaching and learning. Despite expanded offerings in higher education, with the exception of a small number of charter and magnet schools, teaching about the environment and using the natural world as a classroom at the K–12 level has, at best, not been expanding. Although it does not represent a direct measure of classroom time spent on environmental education, as districts have increased their emphasis on English language arts and math, there has been more than a 30 percent decrease in time spent on science and history/social science, key areas of knowledge students need in order to understand the environment.[5]

NEW MODELS FOR STUDYING THE ENVIRONMENT

Over the past eighteen years, nongovernmental organizations, universities, and state agencies have focused on reversing this trend by working with educators to develop and demonstrate how standards-based education can be implemented in concert with developing knowledge about the environment. In 1995, a group of twelve states' departments of education, with the support of The Pew Charitable Trusts, joined together to found the State Education and Environment Roundtable (SEER). These agencies created SEER with the intention of constructing strategies to strengthen student achievement in academic content standards by using high-quality environmental education as a centerpiece for classroom instruction. Since then, SEER has consulted with over 700 schools, districts, and the State of California to create a new form of environmental education called "EBE," or environment-based education—education focused on students learning about the interactions between natural and human social systems. Our work at SEER has resulted in a growing body of research and examples of instructional models that demonstrate how to simultaneously achieve student proficiency with standards while developing their understanding of how Earth's natural systems have been affected by human economies and cultures, and vice versa.

As SEER's director, I have been leading this work since the organization was founded, traveling the country working with educators at hundreds of schools and education agencies in nineteen states. I decided to write this book because the major educational and environmental challenges that our society is currently facing are inextricably connected to the ways humans interact with the world around them. Going back to 1981 when I began working with the government of Costa Rica to create an environmental education program for all of their primary schools, and more recently through all my work with SEER, I have reached the conclusion that changing the way teachers teach and students learn is the only way to develop an educated citizenry capable of resolving these challenges. Even more important, I have had the opportunity to see uncountable numbers of teachers and students succeed with programs like this—fourth-grade students learning about legislative processes so that they could protect Georgia's tree frogs; sixth-grade students cleaning up a polluted creek in Pennsylvania while learning civic responsibility and improving their community's health; eighth-grade students in Massachusetts analyzing data about sewage discharges into Buzzards Bay to determine compliance with federal standards; and high school students in California working on their school district's recycling and waste diversion program as they studied integrated science, math, and technology. Their stories, and those of many others who created EBE programs throughout the United States, are featured in this book.

One of the key reasons that SEER has had success working with teachers and administrators in schools, school districts, and state departments of education across the United States is that EBE programs are focused on students learning about the environment (particularly the interactions between natural and human social systems), about how civil societies and democracy work, and about the variety of perspectives that are represented in their communities—EBE is not intended to turn students into political activists. Occasionally, however, the students' community-based investigations may result in efforts to meet with and inform lawmakers and landowners to mitigate negative environmental impacts that they have documented. They would do this only after taking into account all relevant facts based on understanding an environmental issue and what factors can drive changes in opinions, policies, and laws in democratic systems.

ABOUT THIS BOOK

This book explores how students can master the content and skills in a broad range of academic standards while developing their knowledge about the environment. Using many examples of schools, districts, and the State of California—the only state to adopt an EBE curriculum—it will show how EBE can effectively be used as a context for teaching and learning English language arts, math, science, and history/social science. The programs described throughout will also illustrate how this approach can be effectively implemented in traditional public schools, magnet schools, and charter schools at the classroom, school, district, and even state level.

This book can be used as a guide by teachers, school leaders, and policy makers to plan, develop, and successfully employ environment-based education to teach subject-area standards while students are studying in and about the environment. Since the environment is the starting point for all EBE programs, wherever a school is situated, whether in a rural area, suburbia, or an inner city, educators can successfully design and implement programs that are effective because they are based on locales and content that have meaning for their students.

The book is divided into two major sections. The first section discusses why EBE is important, how it was developed, and how it can directly support standards-based instruction. The second section delves into the nitty-gritty of putting together an EBE program. Whether it is a classroom curriculum unit for a single classroom or a districtwide or statewide initiative, the process is similar. The chapters in this second section cover the important steps of planning and implementing a program, including: developing a statement of mission, vision, and goals; selecting an appropriate and instructionally effective environmental context; targeting a group of academic standards and learning objectives; designing and developing instructional materials; and finally, examining key aspects of student assessment and program evaluation. In addition, the second section and appendixes present examples from several successful EBE programs that can be used as a starting point for creating an EBE program to meet a school's, district's, or state's specific needs.

Over the next decade, standards-based education will be undergoing substantial changes in the wake of the development and widespread adoption of the Common Core State Standards. And, as of this writing, a next generation of science standards is being vetted at the national level. The tidal wave of instructional change that will come with this shifting landscape of standards will oblige states and school districts to undertake a major reworking of everything from textbooks to professional development for teachers—in-service as well as for those just entering the field. This massive shift in K–12 education, when considered in light of the public's rapidly growing awareness of and concerns about the environment, makes it the ideal time to consider using the environment as a context for standards-based learning and for developing critical-thinking and other skills. While EBE is not a magic bullet that will solve all of the world's problems, the benefits that it has to offer are worth considering at this important time.

PART ONE

Introducing Environment-Based Education

CHAPTER ONE

Why the Environment Belongs in Today's Classrooms

Whether in newspapers, on television, or on the Internet, almost every day we hear two themes: one a growing litany of environmental problems, and another reminding us of the educational challenges facing our communities, states, and nation. It is not mere happenstance that we are hearing about these issues simultaneously, for, ultimately, both education and the health of the environment are related to the growth of human societies and the many human-driven changes that have taken place across Earth over the past few centuries.

During this time, we have achieved many great innovations in science, manufacturing, and communications and have also dramatically increased our understanding of how the natural world functions. However, this new knowledge has made one thing abundantly clear—the challenges faced by human civilization and the natural environment are deeply interwoven. And this complexity provides a myriad of opportunities for student learning.

The Ohio teacher who once said to me, "But we don't have an environment at our school," had it all wrong. She was ignoring the fact that we all live in and depend on the environment, whether we are at home, in school, working on a farm or in a factory, driving down the street, growing vegetables in our backyard or buying them in a market, or walking along a trail.

The difference among these settings is that each of them depends upon and involves interaction with different environmental processes and components. Whether we grow our own vegetables or buy them, we depend on: uncontaminated good soil and the decomposers that release nutrients into the soil; the availability of clean water from the water cycle or a water treatment plant; and energy to transport the seeds to the garden shop or the produce to the store. Whether bacteria, fungus, plant, butterfly, fish, salamander, alligator, eagle, or human, all organisms depend on the natural world to meet their needs for survival. The dependence of organisms on the environment represents a complex web of interactions, interconnections, and interdependence.

In the process of getting what they need to survive, organisms often significantly affect the same environment on which they depend. Even microscopic organisms can change their environment. For example, during the natural chemical process we call fermentation, yeast produce carbon dioxide and alcohol. When they are confined to a small space, like a beer bottle, yeast can change the chemistry of their own environment so much that they die from alcohol poisoning.

The effects on the environment of somewhat larger and more complex organisms are often more visible to the human eye—as when thousands of leaf-cutter ants strip tropical trees of their leaves. Larger, more highly evolved animals, such as beavers, change more extensive areas of the landscape and in a wider variety of ways—cutting down trees, diverting waterways, and flooding areas that would otherwise be dry.

THINKING ABOUT SYSTEMS

The natural and designed world is complex; it is too large and complicated to investigate and comprehend all at once.

—National Research Council,
National Science Education Standards[1]

Earth's natural systems and human social systems are collections of interrelated parts and processes that interact with each other and function as a distinguishable whole. The term "natural system" refers to the interacting components, processes, and cycles within a "natural" environment and the interactions among organisms and their environments. Similarly, the phrase "human social systems" encompasses individual humans and human societies, and their political, social, cultural, economic, and legal systems.

Systems thinking is an approach that can be used to investigate any complex system in a comprehensive manner—seeking to understand its structure, the interconnections among all its component parts, and how changes in any element can affect the whole system and its constituent parts over time. This holds true whether the system is a natural system composed of organisms and their natural surroundings or a human social system such as transportation and education systems.

The component parts of systems vary widely according to the type of system. In a natural system, the components might include living things such as plants, animals, and bacteria, as well as nonliving substances such as soil, water, and rocks. The parts of a human social system include individuals, groups of people, businesses, government agencies, legislative bodies, philosophies, rules, television news, the Internet, railroads, and highways.

In addition to their component parts, systems include processes and interactions essential to the functioning of the system as a whole. Predator-prey relationships, energy flow, fermentation, decomposition, precipitation, evaporation, and soil erosion exemplify some functions of natural systems. Similarly, human social systems are dependent on processes like communication, interstate commerce, legislation, law enforcement, wastewater treatment, and economic development.

Teaching students to identify and to think in terms of systems is important to developing their understanding of how the world around them functions. This approach, a central goal of the type of environment-based education detailed in this book, helps students learn how to look at the big picture rather than just the individual components and isolated processes found within a particular system. Further, this type of analysis allows them to consider systems from the perspectives of individuals, human communities, and society as well as to take into account the sustainability of Earth's natural systems.

POPULATION GROWTH AND THE NATURAL WORLD

Human actions have a direct effect on the environment. Our ancestors, like modern humans, depended on their natural surroundings for food, water, energy, and shelter. As humans evolved, they became better adapted to their environment and more adept at gathering, hunting, and fishing to meet their survival needs. The anthropological and fossil records provide evidence that, over three million years ago, some of our earliest ancestors in Africa began using tools to kill and carve up large animals. The development of these tools presents the earliest available evidence of humans beginning to change their local environment, in this case by hunting and killing wildlife.

With their relatively small populations and unsophisticated tools and skills, early humans moved frequently, depleting the resources of one area and moving on to another. This impact, however, was generally small and benign enough to allow these geographical areas to recover. For example, hunting and killing a few animals or collecting berries and nuts when they were in season typically did not cause long-term damage to the places where they lived. Even a few hundred years ago, the fields around the city of Paris were able to process the human waste created by the city's population of a few hundred thousand.

Over many millennia, as humans' tools, skills, and knowledge grew, their effects on the environment continued to increase. With their geographical expansion out of Africa into Asia and Europe beginning about 1.7 million years ago, humans have had a greater and greater impact on the planet. Earth's human population continues to grow exponentially. In 1500, our planet supported about 450 million people; by 1900, the human population had grown to 1.6 billion. Now, just 110 years later, it exceeds 7.1 billion. The increase from 1900 to 2010 represents growth of over 425 percent. Figure 1.1 provides a graphical representation of the growth of the world's human population from prehistoric times to the future.

Increased demands for natural resources have accompanied the rapidly growing human population and, in many cases, have grown faster than the population. As one small illustration of this skyrocketing growth, demand for pig iron, one of the main components of the steel used in industry, increased from 66.6 million metric tons in 1910 to 1,240 million metric tons in 2009—larger by a factor of almost 20, compared to the 4.4-factor increase in human population growth.[2]

Raw materials used in manufacturing are not the only natural resources that the growing human population is consuming at a rapid rate. Demands for drinking water and freshwater for agricultural use have also grown more rapidly than the human population. Water withdrawal during the twentieth century grew twice as fast as the human population. Equally important, demands for water over the next two decades are projected to double from the current rate.[3]

FIGURE 1.1

World population growth through history

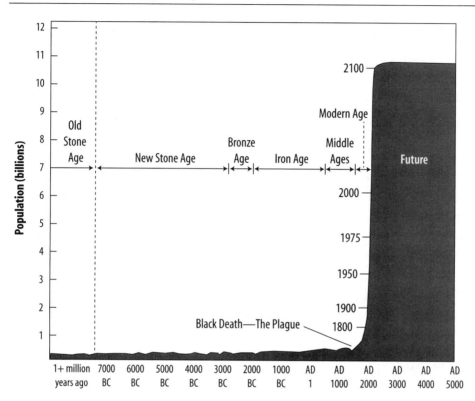

Source: Population Reference Bureau; and United Nations, *World Population Projections to 2100* (1988). © 2006 Population Reference Bureau

As demand for natural resources has grown with the human population, farmers, miners, providers of water, and other industries have expanded operations to meet their community's needs. They have developed new approaches to agriculture as well as more sophisticated tools, equipment, and processes for gathering resources from the environment and transporting them to human population centers. These innovative methods come with their own new challenges and costs. For example, genetically modified crops like soybeans, cotton, corn, and rice can pose threats to native plant species and ecosystems by introducing new genetic material by cross-pollination, thereby outcompeting local organisms and changing an area's biological diversity.

Human impact on the natural environment is not due to just one of these factors; it is the result of all of them combined. Further, the environmental changes produced are cumulative. Overharvesting a forest, for example, results in soil erosion, which in turn pollutes rivers that then may no longer have the capacity to support the plants and insects required for survival of the local fish population. The decreased fish population can then no longer support the economy of a local fishing village.

Environmental impacts of population growth affect us all, whether we live in cities in industrialized nations or in the tropical rain forests in a developing country. Which problems we are most aware of and affected by depend on many factors, including where we live, our economic status, what we do for a living, our individual health, and even our recreational pursuits. Table 1.1 provides examples of some of the environmental concerns found in urban, suburban, rural, and unsettled areas.

Whatever the circumstances, the fundamental cause of all environmental problems is an ever-increasing demand for natural resources—including food, water, energy, shelter, and places to live—for a human population that is growing at a rate of about 1.1 percent per year, or 210,000 people per day.

ADVERSE IMPACTS

All of the methods and strategies we employ to meet these demands for food, water, energy, shelter, and places to live affect different aspects of the natural world and result in wide-ranging environmental impacts that can adversely affect human and/or natural systems. While there are many ways to describe these impacts, they can generally be grouped into seven major categories:

- Air pollution
- Energy production and consumption
- Global climate change
- Loss of biological diversity (biodiversity)
- Water quality and supply
- Ocean degradation
- Overconsumption of natural resources

TABLE 1.1

Environmental concerns in different settings

Setting	*Examples of environmental issues*
Urban	• Air quality (pollution) • Water quality (pollution) and quantity (supply, availability) • Waste management (disposal sites, hazardous materials) • Open space (availability of parks, trails, green space) • Energy (supply, by-products, costs)
Suburban	• Many of the same problems found in urban areas • Urban sprawl (land use, commute times) • Growing population • Energy (consumption, demand, cost) • Natural habitat (loss due to construction, transportation corridors) • Encounters with wildlife (potentially harmful interactions with animals displaced by construction)
Rural	• Some of the same problems found in urban and suburban areas • Pesticides and herbicides (pollution, health effects) • Irrigation water (pollution, supply, availability) • Farmland (loss to urban sprawl, increased cost) • Soil (erosion, loss of nutrients, contamination)
Unsettled areas	• Natural habitat (loss due to resource extraction, including timber cutting, mining, damming of rivers) • Biological diversity (loss of species, changes to ecosystem functions)

Air Pollution

Air pollution is the contamination of the atmosphere with particulate and gas pollutants. Most air pollution is generated in urban and suburban areas, but even rural areas can be affected by large quantities of particulates from activities such as mining and even the plowing of farmland. In general, air pollutants are found in the atmospheric layers nearest the surface of Earth, and the contaminants in these layers, such as ground-level ozone, are the ones that usually have the greatest effects on human health. Atmospheric pollution is often referred to as transboundary, because the constant movement of air causes some of these pollutants to travel great distances around the globe.

Human health is directly affected by air pollution, which has been shown, for example, to cause large increases in the number of asthma attacks in schoolchildren and senior citizens. Asthma affects approximately 10 percent of children in the United States, and the American Association of School Administrators reports that it is the leading health cause of school absenteeism.[4]

Air pollution also affects forests and agricultural crops by reducing growth rates and harming plant survival by increasing susceptibility to disease, pests, and other environmental stressors. The contaminants that make up air pollution can also cause problems for wildlife by compromising their respiration and altering their habitats, as by raising the acidity level in lakes and streams.

Energy Production and Consumption

No matter the type of fuel or its source, energy production (extraction, collection, conversion, transportation, and consumption) pollutes air and water and affects the landscape. Obtaining the raw materials—whether by strip-mining for coal, mining for nuclear materials, drilling for oil and natural gas, or even gathering wood for fuel—has major impacts on the natural world.[5] In Nigeria, for example, the deforestation that results from wood gathering amounts to approximately 1,500 square miles per year.

Collection of energy, even from renewable sources such as solar, wind, and hydroelectric power, significantly affects habitats, plants, and animals when large areas are covered with photovoltaic panels, giant wind turbines, and reservoirs that change river flow and salmon breeding grounds, for example. The next stage of the process, converting those energy resources into forms useful to humans, such as by refining crude oil into gasoline, also results in the pollution of air, land, freshwater, and oceans with materials that range from toxic substances to water hot enough to kill plants and animals.

Transportation of energy resources from the original source to the end user requires equipment that ranges from high-tension cables to trucks, trains, boats, and gas stations—each of which, in turn, uses large quantities of energy and also releases additional air and water pollutants and takes up large tracts of land for transportation corridors. The consumption of energy for everything from driving children to school, to shopping, cooking, and heating homes produces its own array of by-products that have direct effects on the environment and human health.

Over the past several decades, the scope and scale of energy production has been so large that environmental problems are visible at a global level. Energy production and consumption are the largest sources of greenhouse gases in the United States and the principal cause of global climate change.[6] The recent record-breaking episodes of air pollution in Beijing, China, are an example of the pollutants resulting from energy consumption becoming a serious human health hazard, in this case from cars, trucks, and other vehicles.[7]

Global Climate Change

Climate change refers to short- and long-term shifts in the Earth's climatic patterns. Scientists have shown that the Earth's average temperatures have been rising over the past 100 years.[8] This warming is causing a range of changes, including reduced mountain snowpack (a key source of freshwater in places like California), melting glaciers in the Arctic, rising sea levels, and dangerously unstable weather patterns, including widespread droughts and even the amplification of the impact of hurricanes, such as that seen in the flooding from Superstorm Sandy in 2012.[9]

Scientists believe global climate change is caused by industrial development, energy consumption, and automobiles releasing greenhouse gases like carbon dioxide into the atmosphere. These gases trapped in the atmosphere produce a greenhouse warming effect by absorbing infrared and thermal radiation in the Earth's atmosphere and reflecting it back to the surface.

Major changes to the global climate are already having significant effects on natural ecosystems around the world and could, in the long term, influence human societies and populations. For example, as a result of long-term climatic changes, the savanna and forest ecosystems of northwestern Africa, which once supported the Kiffian and Tenerian cultures, are now covered by the Sahara Desert.

Loss of Biological Diversity

Biodiversity loss occurs when plant and animal species die out due to habitat modification, destruction, pollution, or overconsumption as a result of harvesting, hunting, or fishing. Biological diversity—a measure of species' richness, ecosystem complexity, and genetic variation—varies significantly among Earth's ecosystems, ranging from the most diverse tropical rain forests to the least diverse polar regions. Some ecosystems, such as tropical rain forests, are particularly susceptible to the loss of species that live in specialized habitats. Other ecosystems have been transformed by human use into areas that can no longer sustain their native animal and plant populations.

During the past 100 years, for example, 99 percent of the native grasslands in California's Great Central Valley have been plowed and transformed into farmland.[10] Urban sprawl, a growing concern over the past 50 years, is yet another process that converts natural ecosystems into paved and planted suburbs. The resulting species loss diminishes the gene pool and the variety of species needed to maintain diverse and stable ecosystems—diversity needed for humans to develop everything from new farm crops to pharmaceutical products. (Chemicals from native plants and animals provide the source materials or patterns for about 57 percent of the top 150 prescription drugs used in the United States today.[11])

Water Quality and Supply

Water quality and supply are directly affected by natural events and human activity, including climate change. An adequate supply of clean freshwater is vital to human health and survival. Humans, like most living things, require water for the functioning of basic bodily processes including cell division, respiration, digestion, metabolism, and perspiration. They also need large quantities of water for activities such as farming, manufacturing, generating electricity, preparing food, transportation, and recreation. (According to the United Nations, each person needs 20–50 liters of water per day for drinking, cooking, and cleaning. It takes approximately 2,000–5,000 liters of water to grow and process the food to meet each person's daily needs.[12])

In addition to using large amounts of water, humans have an impact on water quality through a variety of activities, including agricultural use of pesticides and fertilizers, and soil erosion; resource extraction and manufacturing processes that release toxic substances; and production and disposal of large quantities of human waste—about two million tons of which are disposed of in water courses around the world every day. (The United Nations Environment Program reports that over half of the world's hospital beds are occupied by people suffering from illnesses linked to contaminated water.[13])

Degradation of water quality and supply can also have substantial effects on natural systems. Chemicals such as the fertilizers and herbicides sprayed on farmland frequently enter streams and lakes, killing plants and animals, accumulating in the food chain, and even reaching the human food supply. A significantly decreased quantity of water flowing in a freshwater system as a result of dams, for example, damages and can permanently destroy habitat for fish and other aquatic animals.

Ocean Degradation

Over the past two decades, water pollution, acid rain, lowered oxygen levels, overfishing, expanding "dead zones," and the effects of global climate change have become major threats to the ocean ecosystem. The overexploitation of oceanic fish stock (which about one billion humans depend on as their principal source of protein) has increased rapidly in recent years, reaching 32 percent of global stocks.[14] Overfishing cod around Georges Bank off the coast of Massachusetts, for example, has led to a 90 percent lower catch over the past thirty years, according to the National Marine Fisheries Service.[15]

Approximately 50 percent of coastal wetlands in the continental United States have been destroyed over the past two centuries.[16] This devastation has eliminated the habitats that once served as the nursery areas for salmon, striped bass, lobsters, shrimp, oysters, and crabs, as well as the breeding grounds for tens of thousands of coastal birds.

Many of the ocean's pollution problems start on land as agricultural by-products, industrial waste, and litter that are carried downstream by runoff water. The "Pacific Garbage Patch" is a global-scale example of this problem. This island of trash that floats in the North Pacific Gyre is estimated to be as large as five million square miles.[17]

Overconsumption of Natural Resources

Natural resources (timber, soil, plants, animals, water, minerals, energy) are typically grouped into three major categories: inexhaustible, renewable, and nonrenewable. Energy from the sun is an example of an inexhaustible resource that will, for all intents and purposes, never run out. Trees and water typify renewable resources, since, if properly managed, they are naturally replenished at a rate comparable to the rate at which humans consume them. It is nonrenewable resources, such as petroleum and natural gas, that are not naturally replenished when they are consumed and are therefore finite and exhaustible.

Earth's growing human population, accompanied as it is by the industrializing economies of countries like China and India, is resulting in rapidly growing global demands for renewable and nonrenewable resources. Accommodating these growing needs using current practices will produce an equally rapid growth of environmental problems, including those associated with agriculture, mining, and manufacturing—the release of pesticides and the production of waste rock and tailings piles, oil and chemical spills, and exhaust gases.

Maintaining the functions and health of Earth's natural systems in the context of the growing human population will require significant changes to some human practices, including some that have been widely used for centuries. In order to avoid damage to natural systems, rates at which resources are being harvested must be adjusted to accommodate the natural replenishment of the resources—overfishing, overhunting, or clear-cutting of forests can, for example, cause long-term damage to an ecosystem and may even lead to the extinction of species.

Table 1.2 identifies the environmental issues associated with fulfilling human demands for each of the five principal resources needed for survival. Little has

TABLE 1.2

Environmental issues and human demands

	Environmental issues						
Resources needed by humans	Air pollution	Energy production and consumption	Global climate change	Loss of biological diversity	Water quality and supply	Ocean degradation	Overconsumption of natural resources
Food	X	X	X	X	X	X	X
Water		X		X	X	X	X
Energy	X	X	X	X	X	X	X
Shelter	X			X	X		X
Places to live	X	X	X	X	X		X

been done to develop environmentally sustainable practices that create a balance between human consumption of resources and the long-term functioning of natural systems. Most efforts to date have been focused on reducing consumption and on recycling or repurposing everything from electronic products, toxic and hazardous materials, leftover food, compostable organic materials, building supplies, and packaging materials. While "reducing, reusing, and recycling," makes a popular catchphrase, it does not represent solutions to the problem of overconsumption of natural resources. Those "solutions" do not resolve the problems associated with harvesting, extracting, transporting, and consuming natural resources.

Other Environmental Impacts

In addition to those that fall within the seven major categories just described, there are numerous other problematic environmental conditions that arise from the interactions among natural and human social systems, including noise pollution, complications from genetic engineering, the introduction of nonnative species, pest management, stormwater runoff, the clear-cutting of forests, and soil erosion.

Decisions regarding where to site waste disposal and treatment facilities or manufacturing facilities, for example, have historically been based on the costs associated with purchasing the land and, in some cases, the possibility of members of the local community protesting the project and initiating a legal battle. The factors associated with these decisions exemplify concerns for environmental justice—the concerns associated with implementing environmental protection programs and policies in a way that promotes equity and fair treatment for all, regardless of race, age, culture, economic status, or geographic location.

ENVIRONMENTAL ISSUES AND SYSTEMS THINKING

All the environmental problems described are the result of a wide variety of interactions between humans and the natural world. These concerns, whether large or small, are complex because they are the result of interactions among large numbers of living and nonliving "moving parts." Environmental issues can, therefore, only be understood if considered from a "big picture" perspective, so providing a platform for learning about systems thinking is another important benefit of bringing the environment into classroom-based education.

Take, for example, the complex interactions leading to the loss of biological diversity resulting from the effects of coastal development (roads, stores, housing, recreational facilities) on a nearby coastal wetland. To fully understand and analyze a situation such as this requires an examination of the interactions between associated natural systems and human social systems. In this case, systems thinking provides a method for evaluating the coastal wetland itself, the plans for a coastal development, the likely effects of coastal development on the wetland, and the potential use of the coastal wetland area on coastal development.

The coastal wetland and the planned coastal development activities each encompass a variety of components, processes, and interactions. Figure 1.2 presents a small set of examples of the components, processes, and interactions found in coastal wetlands, including the plants and animals and interactions among them, and some natural processes and cycles. Figure 1.3 summarizes just a few of the human social systems that come into play if an individual, community, business, or government agency is "developing" a coastal wetland for human activities.

Examining the natural system by itself allows students the opportunity to begin understanding how organisms and ecosystems function—preparing them with an important base of scientific knowledge. Similarly, studying the political, economic, and legal systems of wetland development enhances their understanding of these fundamental aspects of human societies. In the vast majority of K–12 schools,

FIGURE 1.2

Components, processes, and interactions in a coastal wetland

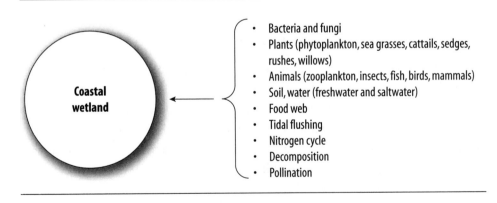

- Bacteria and fungi
- Plants (phytoplankton, sea grasses, cattails, sedges, rushes, willows)
- Animals (zooplankton, insects, fish, birds, mammals)
- Soil, water (freshwater and saltwater)
- Food web
- Tidal flushing
- Nitrogen cycle
- Decomposition
- Pollination

FIGURE 1.3

Components, processes, and interactions in coastal development

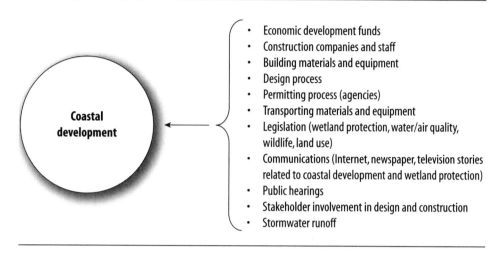

- Economic development funds
- Construction companies and staff
- Building materials and equipment
- Design process
- Permitting process (agencies)
- Transporting materials and equipment
- Legislation (wetland protection, water/air quality, wildlife, land use)
- Communications (Internet, newspaper, television stories related to coastal development and wetland protection)
- Public hearings
- Stakeholder involvement in design and construction
- Stormwater runoff

content related to natural systems (life sciences, earth sciences, biology) are taught separately from subject matter related to human social systems (history, cultures, politics, economics, laws, and geography). As a result, traditional classroom practices do not offer students opportunities to learn about the interactions between natural systems and human social systems.

Figure 1.4 illustrates the interactions that could occur in a development project in or near a coastal wetland. The list below provides examples of just a few of these interactions:

- Construction of buildings and roads decreases wetland habitat, affecting the breeding grounds and activities of fish, birds, mammals, and many other organisms.
- Construction disrupts the natural flow of water in a wetland and the tidal flushing through it, causing flooding that damages businesses, residential areas, and roads.
- Construction noise and the operation of roads, stores, housing, and recreational facilities degrade plants and wildlife, as well as the aesthetic character of the wetland.

FIGURE 1.4
Interactions between a coastal wetland and a coastal development project

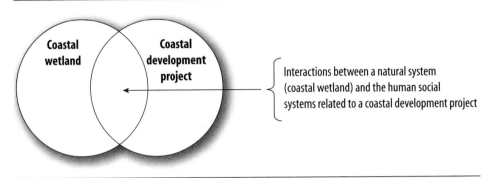

- Impermeable surfaces (roads, parking lots, sidewalks, buildings) decrease the sediments and nutrients entering the wetland.
- Air and water pollutants from highway traffic, runoff from parking areas, and the presence of houses affect remaining wetland and groundwater.
- Buildable land created from the wetland allows for economic development activities (stores, housing, recreational facilities, jobs).
- Access to natural wetland habitat attracts tourism and recreation activities.

Developing an understanding of the interactions between natural systems and human social systems prepares students to become active participants in their communities and a civil society, capable of making well-informed decisions that take into account the potential impacts of human activities on the environment. Equally important, it helps students understand how humans and human culture can be affected by a changing physical world—in all likelihood something that twenty-first-century students will need to know, given the magnitude of some of the environmental impacts described above. In the next chapter, we will explore the evolution of environmental education and how EBE (environment-based education), in particular, is being used to teach systems thinking as an approach to learning about the environment and to contemplating solutions to environmental challenges.

CHAPTER TWO

Standards and the Making of EBE

Throughout most of history, children learned about the environment naturally. Working and playing alongside their families, they learned about the natural world around them just by participating in the business of life. As they grew, children had the chance to watch their parents hunt, fish, make tools, start fires, and prepare food. With the dawn of the agricultural age, they observed and joined their parents in plowing, planting, irrigating, and harvesting. In the fields, forests, deserts, and mountains, they had the chance to watch predators chasing and capturing prey. As their communities grew and the knowledge base became more sophisticated, both adults and children began to see and learn the consequences of catching too many fish, taking too many deer, and depleting the soil. All of these experiences gave children the opportunity to observe and learn about the interrelationships, connections, and interdependencies between themselves, their families, their tribes, and the natural world.

Over time, humans developed a wide variety of approaches to transferring knowledge and skills to subsequent generations. At first, children learned by watching. Recent work in European caves has shown that many children worked alongside the adults, and even recorded what they knew about animals and plants on cave walls themselves. For many millennia, these pictographs were a record of the education children were receiving from real-world experiences about how they depended on the natural world for their survival. With the development of cuneiform script in the ancient civilization of Sumer (located in what became southern Mesopotamia), people began to make records on clay tablets. This was followed by the ancient Egyptians' development of hieroglyphics, a more complex system of writing. Eventually tablets became scrolls, scrolls became books, and books returned to tablet form—only electronic this time.

During the last 150 years, most educational experience has been bounded by the four walls of the classroom, with the gap between children and their environment growing, especially over the past 100 years. At home, fewer and fewer children now spend time exploring ponds, woods, and fields—opting instead for televisions, computers, and video games—while at school, work with textbooks, workbooks, computers, and videos replaces more active learning opportunities like field trips and hands-on activities like planting gardens, studying wildlife, and raising butterflies.[1]

There is no need for this separation between classroom-based learning and opportunities for students to learn in and about the environment—even in an educational system driven by state standards and standardized assessments. Quite the contrary, diverse benefits can be achieved by well-conceived instruction that uses study of the environment as both content and a context for standards-based classroom instruction.

FORMAL STUDY OF THE ENVIRONMENT

There is an extensive historical record of human efforts to learn about the natural world. This started with Thales of Miletus (ca. 620–546 BCE), whom Aristotle identified as the first philosopher and the founder of natural philosophy, the precursor to what we now call science. In its earliest form, the field of biology—the study of life and living organisms—focused principally on natural history and the naming and describing of organisms and their behaviors. Pliny the Elder (23–79 CE) was a dedicated observer of the environment, recording his first-century observations of the world in thirty-seven volumes of *Natural History*. In the eighteenth century, Carl Linnaeus, a physician, botanist, and zoologist, developed the system of binomial nomenclature still used today for naming and classifying all living things. In the following years, scientists like Charles Darwin, Alfred Russel Wallace, and John James Audubon expanded upon and corrected earlier investigators' observations of the natural histories of living things.

In the late nineteenth century, scientists began to develop a new field of study called ecology. This science represented the beginning of a major transition in biology, from descriptive approaches such as natural history, to the study of ecosystems—the relationships, interactions, and interdependence of organisms with their environment and each other. By the 1960s, the growth of our knowledge about the impact of humans on Earth's natural systems led to the pioneering work of Rachel Carson on the widespread harm caused by the pesticide DDT.

A Growing Need for Education About the Environment

With such a rapidly increasing base of knowledge, of necessity, people developed new approaches to teaching children how to survive in an ever-changing world. During the Paleolithic era, it had sufficed to teach children how to make tools and to fish and hunt. With the development of agriculture came the need to pass along knowledge about growing plants and husbanding animals, in addition to information about tools and techniques like the shaduf and saqiya developed by ancient Egyptians to improve irrigation. Around the fifth century BCE, the ancient Greeks developed the concept of schools to formally address the need to pass on knowledge. They were followed several centuries later by the Chinese and Persians. Over the next 2,000 years, the number of schools expanded almost exponentially. Today, for example, there are almost 100,000 schools in the United States.

The Industrial Revolution was rapidly accelerating by the mid-1800s and brought with it a shift from rural-agricultural to more urban lifestyles that led to the development of larger and larger school systems and a greater disconnection between children and the environment. Hours that were once spent outside were traded for an increasing amount of time in the classroom. Although many of the books used at the time, including even the earliest McGuffey Readers, contained many stories about nature, some scientists and educators became concerned that this was not enough exposure to generate an awareness of and interest in plants, animals, and other aspects of the surrounding world. Their efforts during the next fifty years represent the first steps toward formally incorporating the study of the natural world into the school curriculum.

Environment-Focused Programs for Schools

In the late 1800s, a movement called "nature study" began to take shape. In his 1891 book *Nature Study for the Common Schools*, Wilbur S. Jackman, a faculty member of the Cook County Normal School, presented the first known plan for implementing nature study in schools. Liberty Hyde Bailey, one of the founders of the movement, wrote about Jackman's plan: "Instead of looking upon nature study as being supplementary to reading, writing, and other forms of expression, nature study in itself [becomes] a demand that [these other] subject(s) should be taught."[2] Bailey then built on that idea, and in 1901 wrote that nature study had as its goal "to open the pupil's mind by direct observation to a knowledge and love of the common things in the child's environment . . . to put the pupil in a sympathetic attitude toward nature for the purpose of increasing the joy of living."[3]

In the early part of the twentieth century, the U.S. Forest Service identified the need for what they called "conservation education." These programs sought to engage the public in the conservation and the wise use of the nation's natural resources. Even today, the U.S. Forest Service defines the role of conservation education as helping "people of all ages understand and appreciate our country's natural resources—and learn how to conserve those resources for future generations." The U.S. Forest Service, like many other organizations with conservation education programs, views their efforts as part of a larger land and resource management strategy.[4] Over the years, a wide variety of government agencies and private organizations have developed conservation education programs for students, but their principal goal is to use education as a tool to improve the management of natural resources rather than to improve education per se.

In the 1920s, a new field called "outdoor education" was promoted as a way to educate students about the natural world and also to teach academic content. This has changed over the past several decades. Outdoor education has become more of an umbrella term that describes a wide variety of outdoor experiences, such as camping education, experiential education, adventure education, and scouting. These programs are typically taught by specialists in camping, nature studies, resource management, adventure skills, and recreation. Some outdoor education programs make an effort to align to classroom content, but most focus on teaching outdoor activities and skills, developing a sense of stewardship of the natural world through appreciation of and understanding nature, and building survival and leadership skills.

By the late 1960s, individuals interested in protecting the health of the natural environment began to promote what they called "environmental education" to engage the general public in resolving environmental problems. The common focus of these programs has been to engender sensitivity around changing consumption and wasteful behaviors. Consequently, the materials, instruction, and professional development that environmental education programs produce generally focus on environmental goals rather than on helping students achieve proficiency in academic achievement or state and district content standards.

More recently, several organizations have begun to use the phrase "place-based education" to signify environmental education programs that are focused on the local community. The Center for Place-Based Education at Antioch University New England, for example, defines place-based education as "learning within the local community of a student. [To provide] learners with a path for becoming active citizens and stewards of the environment and place where they live ..."[5] This approach shares most of the same goals as more traditional environmental education.

The term "green schools" has come into common usage over the past decade. There are numerous individual schools, organizations, and the U.S. Department of Education promoting this broadly defined concept. The focus and definition of green schools varies widely from site to site, but implementation typically targets minimizing the environmental impacts of school buildings (e.g., energy usage), improving students' health, making school facilities more conducive to learning, and, in many cases, connecting science and technology instruction to the facility.[6]

A different but related strategy began in the early 1990s when the College Board established its Advanced Placement Environmental Science course for high school students. This course provides content that students would be expected to learn in an introductory-level environmental science college course. According to the College Board, such coursework is designed "to provide students with the scientific principles, concepts, and methodologies required to understand the interrelationships of the natural world, to identify and analyze environmental problems both natural and human-made, to evaluate the relative risks associated with these problems, and to examine alternative solutions for resolving or preventing them."[7] Course content focuses on scientific topics, such as geology, biology, environmental studies, chemistry, and geography.

STANDARDS, BUT NOT FOR THE ENVIRONMENT

In the United States, the movement toward standards-based education began following the 1983 publication of *A Nation at Risk: The Imperative for Educational*

Reform, a seminal report by the National Commission on Excellence in Education that sounded an alarm that academic achievement in the American educational system was in serious decline. By the mid-1990s, professional education associations and organizations, some with federal funding, had produced an extensive set of documents that identified standards for academic achievement in the fields of English language arts, mathematics, science, history/social science, and geography.

At that time, no national organizations had developed or promoted a comprehensive set of environmental content standards to parallel the other subjects, although the National Geographic Society's geography standards included some topical environmental content. As a result, when most state departments of education went through the process of adapting the national standards to their individual states, and state boards of education adopted their standards, none included more than minimal mention of the environment.

Common Core State Standards

The next major transition in the standards movement began in 2010 with the release of the Common Core State Standards for English language arts and literacy in history/social studies, science, and technical subjects, and mathematics. These standards were developed as part of an initiative that was led by the Council of Chief State School Officers and the National Governors Association Center for Best Practices. The Common Core State Standards are intended to ensure that students receive a high-quality education that is consistent from school to school and state to state. By the end of 2011, these new standards had been adopted by forty-five states, and they will continue to be implemented over the next several years. Like the standards that were developed for English language arts and math during the 1990s, the Common Core Standards do not include content or skills related to the environment.

Pennsylvania and California—Exceptions to the Rule

By 2002, the state boards of education of Pennsylvania and California had adopted standards for English language arts, mathematics, science, and history/social science. Soon thereafter, these two states decided it was necessary to formally incorporate environmental content into their education systems—Pennsylvania in 2002 and California in 2003 as part of its Education and the Environment Initiative (EEI).

In the Commonwealth of Pennsylvania, which has for many years played a leadership role in conserving natural resources and protecting the environment, an interagency group led by staff members from the Pennsylvania Department of Education began to develop academic content standards for "environment and ecology." In 2002, the Pennsylvania State Board of Education approved these as the state's fourth set of standards, following reading, math, and science. Once these environmental standards were approved, the Department of Education began a multiyear process to help administrators of districts and schools develop and implement programs to provide the necessary instruction. More recently, Pennsylvania developed student assessment instruments with an expanded scope that takes into account science, technology, and environment and ecology. Now students in the fourth, eighth, and eleventh grades are tested and expected to be proficient in the environment and ecology content standards.

In 2003, California's legislature and governor passed and signed into law a bill that called for the delineation of a set of environmental principles and concepts that focus on the interactions and interdependence of human societies and natural systems. Since the State Board of Education and the Department of Education had spent several years developing and finalizing academic content standards, the law included a clause that barred these environmental principles and concepts from overlapping or conflicting with existing state standards. The law conceived of these principles and concepts as representing "big ideas" about the environment—critical understandings that every student in the state should know. In addition to requiring the development of the principles and concepts, the legislation called for the creation of a model curriculum designed to teach the principles to students throughout the K–12 system. This curriculum, developed between 2006 and 2010, has been officially approved by California's State Board of Education and is currently being disseminated. Different aspects of California's Education and the Environment Initiative (EEI) are discussed throughout this book.

ORIGINS OF EBE

At about the time that national standards were being developed in the 1990s and states were beginning to formally adopt their individual sets of standards, The Pew

Charitable Trusts funded a study that examined opportunities for integrating the teaching of standards with education about the environment. This 1995 study concluded that few education reformers thought it would be valuable for the environmental education community to develop its own academic content standards.[8] Rather, they suggested that the greatest potential lay with efforts to expand the integration of environmental content into other disciplines and areas of study. This paralleled a recommendation for the worldwide development of environmental education made during the 1977 Tbilisi Conference, which urged the "reorientation and dovetailing of different disciplines and educational experiences which facilitate an integrated perception of the problems of the environment, enabling more rational actions, capable of meeting social needs . . ."[9] The study also concluded that environmental educators did not consider it a high priority to undertake an in-depth integration of environmental content with the key disciplines of English language arts, math, science, and history/social science.

Based on the results of this study, the Trusts provided funding in 1995 for the establishment of the State Education and Environment Roundtable (SEER) as a cooperative endeavor of twelve states' departments of education; this eventually grew to include sixteen states. This San Diego–based consortium of agencies sought to infuse environmental content into the K–12 education system by integrating it with education standards and frameworks, student assessment, and professional development programs for teachers. For the next ten years, these agencies' representatives gathered together for two weeklong conferences each year.

With members of state departments of education around the country in the process of developing and adopting academic standards, SEER's members identified the critical need for research into the academic efficacy of environmental education. The member agencies called for a study of existing research into the efficacy of teaching content standards through environmental education. In response, SEER's staff members examined 600 research studies and publications in the areas of both general and environment-based education. This investigation, *The Educational Efficacy of Environmental Education,* revealed that general education research held little evidence relevant to determining the educational efficacy of environment-based education.[10] Furthermore, it found that, although research about traditional environmental education was substantial, it had been primarily concerned with assessing the development of environmental skills, knowledge, and behavior, and provided little insight into the effect on the overall educational experience of students.

Since the literature survey indicated a need for basic research on the topic, SEER's members asked the staff to design and undertake a study of schools using the environment to teach state and district standards. The study was intended to compile data on the effects of environmental education on student achievement in reading, writing, math, science, and history/social science. SEER published a report of its findings in 1998 entitled *Closing the Achievement Gap: Using the Environment as an Integrating Context for Learning.* Chapter 3 presents an overview of this study and its results.

SEER's study coined a new term, "environment-based education" (EBE), describing it as "a framework for instruction that focuses on standards-based educational results by using the environment and related issues as a context for instruction."[11] Environment-based education has at its core three major goals: helping students achieve success with academic content standards; developing their understanding of interactions between natural and human social systems; and preparing students to be active members of a civil society with the skills they need to identify and resolve environmental challenges.

Since developing the concept of EBE, SEER has created two major programs using this approach: (1) eleven state networks of EIC Model (Environment as an Integrating Context for learning) schools, in cooperation with its member agencies; and (2) the Education and the Environment Initiative (EEI), with the state of California. These two programs demonstrate parallel but different means of achieving the same instructional goals: the EIC Model is characterized by the development of localized EBE instructional programs at a school or district level; the EEI curriculum resembles a more traditional, textbook-based methodology.

The Integrating Context Model

In 1997, based on its national study, SEER developed a strategy for implementing EBE called the EIC Model (Environment as an Integrating Context for learning). Working in cooperation with its state department of education members, SEER created an extensive professional development program that, since 1998, has been used to train teachers in over 700 schools in methods for developing and implementing programs using the environment as a context for teaching and learning.

This model program brings together six key pedagogies:

- integrated interdisciplinary instruction that breaks traditional boundaries between disciplines;
- community-based investigations as learning experiences that offer both minds-on and hands-on experiences through service-learning[12] opportunities;
- collaborative instruction so teachers, parents, students, and community members can together connect instruction and learning;
- learner-centered, constructivist approaches adapted to the needs and unique abilities of individual students;
- combinations of independent and cooperative learning;
- local natural and community surroundings as the "context" for connecting these proven pedagogies.

The EIC Model was designed to be employed at the school or district level with grade-level teaching teams working together to design instructional units that integrate content in several disciplines. As described in depth in the second part of this book, implementation begins with the selection of a local environmental concern to study, such as water quality in a local creek, air pollution at the school resulting from idling buses and cars, the consumption and disposal of a resource like paper, or other local topics that interest the students.

One of the first steps in each investigation is to undertake a preliminary analysis of the natural and human social systems that affect and are affected by the environmental issue to be studied, such as the ones posed by coastal development discussed in chapter 1. Teachers and students complete a community map of the natural and human social systems near their school in order to identify one or more local environmental problems as the focus of their community-based investigations. Discipline-specific standards that can effectively be taught in the context of the community-based investigation are then chosen for interdisciplinary units. Teachers complete an instructional design centered around standards-based learning. Typically the instructional design incorporates standards from multiple disciplines—for example, English language arts, mathematics, science, and history/social science—depending on instructional needs. In the most effective units, teachers combine authentic instructional activities related to their community-based investigations with activities from more traditional materials that they use to develop students' skills, such as word decoding and basic arithmetic.

As students investigate the causes and effects associated with the environmental problem under study, they develop the skills they need to communicate their discoveries and become active members of their communities, able to seek the support they need to resolve local environmental issues. Since teachers design these units to support standards-based instruction, they also incorporate either traditional or performance-based assessments into their plans to ensure students are achieving proficiency.

California's Education and the Environment Initiative Curriculum

In 2006, the State of California engaged SEER to lead the development of the curriculum for its Education and the Environment Initiative (EEI curriculum). The development of this curriculum was to be based on a plan reviewed and approved by the Interagency Model Curriculum Planning Committee, which comprised six of California's key education and environmental agencies. The curriculum was focused on teaching over 125 K–12 standards in the disciplines of science, history/social science, and English language arts.

Collaborating with fifty experienced curriculum writers and environmental experts from universities and nongovernmental organizations, SEER's staff members worked with the state agencies to develop eighty-five units designed to teach the standards by having students examine numerous environmental problems important to California (e.g., salinization in the San Joaquin Valley) as well as topics from around the world (e.g., loss of biological diversity). Prior to final production, these instructional units were reviewed by technical experts, field- and pilot-tested in classrooms across the state, reviewed by California's Curriculum Development and Supplemental Materials Commission, and given a final editing.

In 2010, California's State Board of Education unanimously approved this curriculum for use by teachers across the state. They can now use the EEI curriculum in their classrooms to replace instructional materials contained in their adopted textbooks, which, previous to this, had been the only instructional materials approved for classroom use.

The design of the EEI curriculum is intended to:

- help students achieve proficiency with selected California content standards;
- teach K–12 students the key understandings represented by California's Environmental Principles and Concepts (see appendix A at the back of the book) as a means of preparing them to understand future environmental challenges;
- support students as they develop and apply: science investigation and experimentation skills; history/social science analysis skills; English language arts reading, writing, listening, and speaking skills; and mathematics skills;
- integrate well with textbooks and other instructional materials adopted by California's State Board of Education.

Like the EIC Model instructional units, EEI curriculum units are designed to help students achieve proficiency in specific academic content standards through mastery of a series of learning objectives. The EEI curriculum units are focused solely on California's standards, but with minor modifications such as identifying relevant environmental problems and resource materials, they can be adapted for teaching in other states. The EEI curriculum units provide students with the opportunity to read about and discuss environmental problems, but, unlike EIC Model instructional units, they do not directly involve students in community-based investigations or service-learning activities. While both approaches prepare students to be involved in future environmental decisions, the local experiences built into the EIC Model approach frequently give students the opportunity to work directly with local decision makers.

COMPARING EBE TO STANDARDS-BASED PROGRAMS

Helping students achieve success with academic content standards is the central goal of both traditional standards-based and EBE instruction. The principal difference is that the EBE approach teaches the standards using a cohesive context that connects students' new knowledge and skills to something that they can readily relate to—their local environment and community. Thus, in a classroom using the EEI curriculum, California's sixth-grade students can, for example, connect new information about the agricultural techniques used by the early civilizations of Mesopotamia, Egypt, and Kush—a topic most sixth graders would see as irrelevant to their own, present-day lives—to what they already know about agriculture in California's Great Central Valley.

Only a few states include standards with substantial content related to natural systems, although many include the general concept of ecosystems in their science standards. Even more rarely do state standards require students to learn about the interactions among natural systems and human social systems. Since content about systems is rarely incorporated into state standards, it is extremely sparse in published textbooks and other instructional materials used for science and history/social science. The only way, then, to deeply incorporate systems and systems thinking into classroom instruction is to use these understandings as a context within which to teach other academic content.

Released in April 2013, the newly developed Next Generation Science Standards incorporate systems and systems thinking to a much greater extent than did the 1996 National Science Education Standards. At several grade levels, they call for students to explore the effects of human activities on natural systems. As early as kindergarten, these new performance expectations concentrate on content and skills that will prepare students to "communicate solutions that will reduce the impact of humans on the land, water, air, and/or other living things in the local environment."[13] While at the high school level, they call for students to "construct an explanation based on evidence for how the availability of natural resources, occurrence of natural hazards, and changes in climate have influenced human activity."[14] To the extent that these new science standards are adopted by states, EBE strategies can play an even more significant role by providing students with opportunities to apply the science they are learning to the world around them—making it much more likely that they will become proficient in science.

Environment-based education is designed to help students meet proficiency levels of state and school district standards, including the Common Core State Standards, through a school's standards-based academic programs in English language arts, math, science, and history/social science, as well as in other disciplines such as technology, art, and music. It is not a separate topic with its own discrete academic content.

It is not necessary to think of the rich content found within the study of natural and human social systems and environmental topics as additional subjects, or to develop them into new standards. This content is too important to view as separate

Examples of Opportunities for Using the Environment to Teach Standards

English Language Arts

Diverse content related to the environment can be readily found in the form of essays, poems, stories, news articles, opinion pieces, books, technical papers, and at a multitude of Web sites. This material offers content that educators can use to provide instruction in reading, writing, speaking, listening, and language. At the same time, teaching through this content helps students achieve what the Common Core State Standards for English language arts refer to as "literacy in history/social science, science, and technical subjects." The ability to cite and support their analysis of textual evidence, for example, can be developed by sixth graders examining the effects on the agricultural economy of changes in local weather, or by eleventh graders analyzing primary and secondary sources as they explore the diverse points of view about development of the Arctic National Wildlife Refuge.

Math

The Common Core State Standards for math are focused on "standards of problem solving, reasoning and proof, communication, representation, and connections." Studying math through analysis of environmental topics such as the dilution of a pollutant or its distribution in a local creek and river system helps students see and value content such as math operations, statistics, and geometry by letting them experience the real-world benefits of math. Using environmental problems as a basis for math instruction helps students make sense of math because their fascination with the world around them drives them to solve these problems. Sixth graders can learn how to represent and analyze quantitative relationships between dependent variables (amount of carbon monoxide) and independent variables (numbers of cars and buses) by gathering and graphing data about the vehicles idling in their school's driveway.

Science

Science, at its core, involves study of the living and nonliving components of Earth's natural systems, including the interactions among organisms, natural systems, climate, and nonliving resources. These interactions are the driving force behind the survival and evolution of all living things. With the world as their laboratory, students have a chance to do authentic scientific research analyzing interactions between natural and human social systems. Further, while some state and district standards for science contain content about the components and processes encompassed by natural systems, these usually focus solely on an ecosystems perspective. Very few examine the effects of human social systems on natural systems or how human social systems are influenced by natural systems.

History/Social Science

There are a multitude of opportunities to teach about history, civics, geography, and economics by studying how humans—as individuals, communities, and societies—are influenced by and dependent on natural systems. Examining human connections to the environment is crucial to fully understanding these topics, whether they are: tool making and migration patterns among Paleolithic peoples; the genesis, growth, and end of the ancient Egyptian civilization; agricultural practices of ancient Mesopotamia; the growth and dissolution of the Roman Empire; the developments leading to the Industrial Revolution; the forces behind the American Revolution; the great Irish potato famine; or the important present-day concern about the effect of overharvesting ocean fisheries on different economies around the globe.

subject matter when some of its greatest value comes from using it as a means of interconnecting content across disciplines and demonstrating to students how what they are learning in school can be applied to a real-world setting.

Finally, as is discussed in the next chapter, environment-based instruction provides students with the chance to develop many of the skills they will need for college and/or careers and to become active members of a civil society in the twenty-first century. Learning in an environmental context offers students a wide variety of opportunities to examine real-world issues and related decision-making processes as well as their connections to political systems, legal systems, economic systems, sociocultural systems, and the interplay of these with natural systems. Since many studies of environmental problems spark personal interest in the students, EBE frequently leads them to participate in service-learning activities. These service-learning activities can then reinforce academic content as well as build students' communications skills.

As it develops knowledge about the environment, EBE offers an effective framework for instruction that succeeds in teaching to academic standards and preparing students for the twenty-first century. It also meshes well with many of the instructional objectives stated by the developers of the Common Core State Standards. Hence, EBE can be used as the basis for designing instruction that will help state and school district administrators and teachers implement new or existing content standards over the coming years.

COMPARING EBE TO ENVIRONMENT-FOCUSED PROGRAMS

Environment-based education differs from programs that are focused principally on teaching children about the environment. The differences fall into three main categories: key objectives; connection to school content; and the audience focus.

The key difference between EBE and more traditional education programs about the environment is EBE's emphasis on academic content standards. Some environment-focused programs make an effort to align their material to state standards. However, they are not designed or intended to serve the purpose of teaching standards so that students will achieve proficiency. Within an EBE program, the instructional materials and curriculum are designed specifically for the purpose of helping students master standards-based content in a way that also builds their understanding of the environment and environmental issues. Table 2.1 summarizes the differences between the previously described environment-focused programs and EBE.

While it is important for all students to study natural systems, human social systems, and their interactions, the environment can play a much greater role in standards-based education. Teachers and schools across the nation have discovered that the environment can be an engaging and highly effective medium for student learning in all subjects, from preschool through high school. Because the environment is connected to everything around us—from science to history and social science to literature—it offers an authentic and dynamic context for teachers and students to tie together teaching and learning across the core disciplines. In the next chapter, we will examine how EBE uses these interconnections to benefit students in terms of their academic achievement, engagement in school, and preparation for higher education and employment.

TABLE 2.1

Origins and objectives of environment-focused programs

Programs and origins	*Description*
Nature study (late 1800s)	Key objectives: • opportunities for direct observation of nature • knowledge and appreciation of the environment • sympathetic attitude toward nature *Connection to school content:* used as a supplement to traditional reading and writing content *Audience focus:* students
Conservation education (early 20th century)	Key objectives: • understanding and appreciation of natural resources • improved management of natural resources • conservation of resources for future generations *Connection to school content:* sometimes used to supplement traditional school content *Audience focus:* people of all ages
Outdoor education (1920s)	Key objectives: • knowledge of the natural world • sense of stewardship for the natural environment • survival and leadership skills • learning through real-world experiences *Connection to school content:* sometimes used in a school setting to teach academic content *Audience focus:* students, children outside of school, families, adults
Environmental education (late 1960s)	Key objectives: • environmentally literate citizens who know about the environment and associated problems • awareness of how to solve environmental problems • motivation to work on solving environmental problems • changes to attitudes and behaviors that help avoid environmental problems *Connection to school content:* sometimes used in a school setting in ways that align with teaching of academic content standards *Audience focus:* students, children outside of school, families, general public

(continues)

Programs and origins	Description
Place-based education (mid-1990s)	Key objectives: • knowledge of the local community, including its history, culture, literature, and environment • sense of stewardship toward the community and environment • participation in service-learning activities that encourage individuals to become active citizens *Connection to school content:* sometimes used in a school setting in ways that align with teaching of academic content standards *Audience focus:* students, members of the local community
AP environmental science course (mid-1990s)	Key objectives: • knowledge of scientific content and scientific methods • understanding of interrelationships in the natural world • ability to identify and analyze environmental problems and evaluate environmental risks • capacity to examine alternative solutions to environmental problems *Connection to school content:* college-level environmental science content *Audience focus:* high school students
Environment-based education (1997)	Key objectives: • teaching of academic content standards • understanding of natural and human social systems and their interactions • ability to investigate real-world issues and related decision-making processes • skills needed to identify and resolve environmental issues • participation in service-learning activities to encourage active involvement in a civil society • reinforcement of communications skills • preparation of students for college and/or careers *Connection to school content:* instructional design is driven by academic content standards in English language arts, math, science, and history/social science but may include content from other disciplines *Audience focus:* K–12 students

CHAPTER THREE

The Benefits of EBE

A growing body of research demonstrates the many ways in which students benefit by participating in environment-based education (EBE) programs. Although the focus, depth, and quality of the studies vary, taken as a whole, the research strongly indicates that students benefit from participating in EBE programs in three key areas:

1. increased academic achievement related to standards
2. improved engagement in learning and classroom behavior
3. better preparation for college and careers

There is also a great deal of anecdotal evidence of other benefits of EBE programs, including: students learning about real-world issues, participating in solutions, and developing a sense of environmental stewardship; heightened teacher morale; improved school climate; increased opportunities for community and parental involvement. These benefits accrue to students and teachers alike, as attested to by Louise Hope, a teacher at Pickens Middle School in South Carolina, who said:

> What the EIC Model [Environment as an Integrating Context for learning] has done for us as educators is to provide a focus on integrated instruction and meeting state standards. This kind of teaching keeps our team motivated, it is exciting. We are meeting new people, making community connections and providing meaningful, positive service-learning [opportunities]. And, the best thing of all is that our students, faculty, and community now see our kids as leaders and responsible citizens.

Students, like the teachers, see the differences between their typical classroom work and their learning activities in EBE programs. One student at Concrete

Middle School in Washington State expressed what is at once an important aspect of EBE programs and one of its educational benefits when he said, "It was hard because we had to step out of our comfort zone, but you remember what you learn because you got to be involved and do it."

SUMMARY OF THE RESEARCH

The 1998 publication of the State Education and Environment Roundtable's (SEER) report *Closing the Achievement Gap: Using the Environment as an Integrating Context for Learning* presented the results of this first-ever study of the academic benefits of environmental education.[1] This research began in early 1996 when, with the guidance and support of its department of education members, SEER initiated a three-year study of schools that use the environment to teach their state and district standards.

The research team first surveyed hundreds of educators in schools across the country using a set of in-depth selection criteria including: degree of integration of environmental education across the curriculum; student involvement in projects and problem solving; extent of team teaching; program longevity; and demographic characteristics of the community. Ultimately, they selected forty schools in thirteen states that met all of the research criteria. The schools that were asked to participate in the study represented a cross-section of socioeconomic levels as determined by statistics for free and reduced-price lunch.

Data for the study were gathered in four ways: telephone interviews, written surveys, school visits, and through collection of available standardized assessment data. The team used four separate survey instruments to gather a wide range of data about the programs and the effects of teaching language arts, math, science, and history/social science in an environmental context. They then visited each of the schools and conducted extensive interviews with administrators, teachers, students, and in some cases, parents and community members. Over 650 teachers, administrators, and students participated in the interviews. In the schools where data were available, the research team also collected information on test scores, absenteeism rates, and disciplinary actions, as well as comparable data from classrooms and other programs in the same schools that used traditional instructional practices.

The results of this study indicated that students in schools and classrooms implementing what we now call EBE performed better on standardized measures of academic achievement—particularly in language arts, science, and history/social science—than their peers in traditional classrooms. Further, teachers reported, and the available data corroborated, that there were fewer discipline problems among students who were participating in these environmental programs.

Since the publication of *Closing the Achievement Gap*, many other research projects have examined the effects on students and teachers of implementing EBE programs and have reproduced many of SEER's findings. These other research projects have been conducted by a wide range of investigators, including university faculty members, graduate students, members of state departments of education and of nonprofit organizations, and consultants. These studies included both quantitative and qualitative data collection and analysis. Some of the studies had large sample sizes (hundreds to thousands of students) and compared multiyear data, while others represented much smaller sample sizes in terms of numbers of schools and students participating. The majority of these studies examined schools and programs unrelated to SEER's work, while a few looked at schools for which SEER had provided teachers with professional development in the implementation of the EIC Model.

These results must be considered in the light of the limitations typical of all educational research and some that are specific to EBE programs. High-quality educational research is always limited by factors involving program comparability, equivalence of "treatments" of students or teachers, size and comparability of test populations, and the opportunities to implement long-term studies. Research into EBE programs encounters these same limitations and a few more. The Environment as Integrating Context (EIC) Model, for example, is designed to meet the needs of a specific school and the environment where it is located. In addition, EBE programs at different schools may vary in the standards they address and the assessments they use, in teachers' knowledge of the local environment and community, in participation of experienced community partners, and in the availability of funding and other resources. This variability among schools implementing EBE programs adds substantially to the difficulty of conducting parallel studies across schools.

The summaries that follow weave together research results from a wide range of studies of EBE programs to illustrate the impact of these programs on teaching and learning in core subject areas, on student motivation and behavior, and on other outcomes. More information on these studies and their results is listed in appendix B at the back of the book.

ACADEMIC ACHIEVEMENT RELATED TO STANDARDS

English Language Arts

When students read, write, and speak about topics that interest them, as they do in EBE programs, they are more likely to make an effort to strengthen these skills. Students are engaged by reading about the natural and human social systems found in their communities and regions, and in the world as a whole. Teachers frequently observe that students in EBE programs tend to read more and pursue a greater variety of reading materials when they have an opportunity to learn about plants, animals, and natural and human social systems in their own surroundings. Many report that even students who are otherwise reluctant readers begin to seek out more to read because they want to learn more about a particular aspect of the environment in their community. Since content about local environmental topics is often found in newspapers and on the Internet, teachers have an opportunity

---

Reading and Writing About Erosion

At Twin Lakes Elementary School in El Monte, California, Elaine Chiu, a second-grade teacher, designed a lesson focused on one reading and one writing standard. Students examined questions like "How do student and staff activities affect the soil on our campus?" During this lesson, students: brainstormed what they knew about soil erosion; read about and discussed causes of erosion; observed where erosion occurred on campus; and heard about ways to prevent erosion from local experts from the Home Depot Gardening Team, Mountain View School District Maintenance Department, and El Monte Historical Museum. They demonstrated their developing English language arts skills by writing a small book entitled "The Many Causes of Soil Erosion." Ms. Chiu noted that: "The majority of my students are English language learners, and writing is difficult for them. Real-world, environmental experiences helped them to grasp the language."

to support their students with the most up-to-date information for their research activities and investigations, and in the process help students learn how to evaluate these different sources of information.

Language arts studies with environmental content give students an opportunity to discover the power of writing about the issues of the day and how written and spoken language can be used to influence their communities and the world around them. The variety of content in environmental programs helps students develop proficiency in diverse literary styles—from lab reports to poetry, editorials to expository writing. Many teachers report that students who at other times may not enjoy writing become more engaged and creative when they are working on something they care about and can have an influence on, like a local environmental problem or their own investigations. Creative writing assignments that ask students to describe the connection between water pollution and the public health of their community, for example, can make writing more relevant and fun. Also, research into more general environmental topics like air quality can take students back through history from the present to trace an environmental concern with roots in the past, like the Industrial Revolution.

Widespread interest in the environment creates diverse opportunities for students to develop speaking skills as they make oral and visual presentations. Students in EBE programs frequently make exciting discoveries about the local environment and their community, and their passion about what they learn drives them to want to share their new knowledge through oral and visual presentations.

As teachers develop and strengthen their EBE programs, they find that they can capitalize on interest in the environment to initiate opportunities for students to present their work to other classes, neighboring schools, civic organizations, local agencies, and governing bodies. Opportunities like these strengthen students' speaking skills and confidence as they become the "authorities" on topics that others might not be familiar with. Teachers frequently report that students who are generally reticent to speak before their classmates on other content lose this fear in their enthusiasm about the environmental investigations they are working on.

The evidence from SEER's research and other studies indicates that EBE programs can help students develop and strengthen the depth of their knowledge and a variety of skills in English language arts. The gains in English language arts achievement reported for schools implementing EBE programs range from subtle to dramatic. The examples below are representative of the results from EBE research studies.

- Seventy-six percent of standardized test scores indicated that students in the EBE programs performed better than students in traditional programs ($n = 91$ grade-level student assessment comparisons in eight pairs of California schools).[2]
- Eighty percent of grade-level comparisons demonstrated higher scores on Palmetto Achievement Challenge Tests (PACT) by students in the EIC Model programs ($n = 10$ pairs of classrooms in ten South Carolina schools).[3]
- An average of 11 percent more students in an EBE (integrated) program met Washington State standards on reading compared to students in a traditional program (seven-year study).[4]
- An average of 11.5 percent more students in an EBE (integrated) program met Washington State standards on writing compared to students in a traditional program (seven-year study).[5]

Mathematics

Through hands-on opportunities in EBE programs, math is no longer abstract or irrelevant, as students learn by applying these skills to real-world problems. As they begin to perceive the value of math to their everyday lives and futures, they become more enthusiastic about their studies. One student involved in raising funds to build a nature trail adjacent to Concrete Middle School in Washington State put it this way, "We used our math in writing grants . . . It was good; I can use these skills later in my life."

At Dowling Urban Environmental School in Minneapolis, Principal Jeffrey Raison, initiator of their environmental curriculum, worked with a variety of local partners like the Minnesota Valley Wildlife Refuge to guide the students in studies of the nearby Mississippi River. The students' projects gave them the opportunity to develop and apply their math knowledge to activities such as taking measurements of trees as they restored habitat on the twenty-acre school site, getting them involved in real-world undertakings rather than doing worksheets and working out of a book. Raison conducted a pre- and post-analysis to compare about 350 students' math performance in the second through sixth grades before and after teachers began implementing their environment-based curriculum. He discovered that median scores in math comprehension, as measured on the California Achievement Test, rose substantially over the two-year period between 1990 and 1992: 16 percentile points for low achievers; 13 percentile points for middle achievers; and 7 percentile points for high achievers.[6] This pattern of improving student achievement has continued over the past twenty years—Dowling is ranked in the top 25 percent of Minneapolis schools, with a higher percentage of its students scoring proficient or above in reading and math than the average in other city and state schools. Now, many years after the program began, the school's current principal, Joe Rossow, says, "Our distinction is our environmental involvement, our special education programs, a highly qualified and stable staff, and our unparalleled desire to have fun with learning." It is their environmental program that led the Minnesota Elementary School Principals' Association to recognize Dowling as a 2011–2012 School of Excellence.

When students explore school grounds, their surrounding natural areas, neighborhoods, and the community at large, they discover opportunities to employ a

Using Math to Save Trees

Sixth-grade teacher Ann Au Anderson at Jackson Elementary School in Altadena, California, developed a series of interdisciplinary lessons to teach three mathematics standards while helping students achieve their own goal of reducing paper use and waste on campus. The sixth graders designed a data-gathering questionnaire and convinced the school office manager and custodian to record the amounts and types of paper used and discarded on a weekly basis. During this study, students gathered and analyzed data, using information from the office to calculate the overall costs of purchasing, using, and disposing of these products. Students then presented their results, along with recommendations for ways paper use and disposal could be reduced, to the principal. "Students were able to transfer academic skills to real-life situations, making their learning relevant to their everyday lives," says Anderson. "They also made connections across the curricula, noting that all learning is related somehow. Being able to analyze data and present findings to administrators are lifelong skills gained through the project." Now, looking to the future, Anderson says, "As we approach the adoption of Common Core Standards, [this type of investigation] embodies the skills and concepts that we want to teach our children today: making learning the standards relevant to real-world applications."

range of math skills. Elementary school students calculate the average number of cicadas in their schoolyard study plots; seventh graders calculate the board feet of lumber in the woods behind their campus; high school students conduct a complex statistical analysis of the population dynamics of deer in their county—all these students are developing their emerging math skills in authentic settings. Teaching real-world math in this manner also offers students a variety of ways through which they can help their communities—a further demonstration of the everyday value and application of math skills.

In addition, teachers involved in successful environment-based instructional programs state that when math is taught within the context of the environment and community, students become more invested in the accuracy of their answers, and teachers see fewer nonsense answers on math tests.

The gains in math achievement reported for students in schools implementing these programs vary from subtle to dramatic. The examples below are representative of the results from many EBE research studies:

- Students in 65% of schools with EBE programs scored higher (statistically significant) on Washington Assessment of Student Learning (WASL) math tests than otherwise comparable schools over a five-year study period ($n = 77$ pairs of Washington schools).[7]
- Seventy percent of grade-level comparisons demonstrated higher scores on PACT tests by students in the EIC Model programs ($n = 10$ pairs of classrooms in ten South Carolina schools).[8]
- An average of 10 percent more students in an EBE (integrated) program met Washington State standards on math compared to students in a traditional program (seven-year study).[9]

Science

Students in EBE programs are mastering standards at the same time as they are building their scientific knowledge and learning how to apply scientific skills. These students typically do better than their traditionally educated peers on standardized assessments of science content.

Teachers who use the environment as a context for teaching science quickly discover their students are more excited and enthusiastic about learning than students who are learning from traditional textbooks and lab assignments. As one high school senior from Clay County, Kentucky, put it, "When we got to go outside and really do [science] for ourselves, it made the understanding a lot better. I've seen a lot of students I've gone to school with all my life making better grades now because they understand [science] better." The hands-on approaches typical of EBE programs appear to make it easier for students, over a wide range of ability levels, to improve their performance, appreciate science more, and remember what they have learned.

From the students' perspective, EBE programs make science "real." Students are not just reading about insects, trees, streams, and creeks, they get the chance to become active participants in projects that look at these important components of the ecosystem from a larger perspective. As George Radcliffe, a teacher at Centreville Middle School, Centreville, Maryland, puts it: "There's no better way to study science than to collect and analyze data from your own river. Being part of a community project adds legitimacy and a seriousness of purpose that wasn't there before. We're now part of something bigger."

Examining the complex interactions between natural and human social systems allows teachers to review the skills involved in conducting real scientific studies and

Using Science to Investigate a Mystery

When students at Seven Generations Charter School in Emmaus, Pennsylvania, realized they were not seeing many bees in their neighborhood, they asked their teachers for help in figuring out if this was normal. Fifth-grade teachers Alison Panik and Abby Mahone decided to connect their students' concerns to several science standards through a study of pollinators in their area. They began by examining how their school garden affected the pollinator population in their community. Students built and maintained two beehives on the school campus, designed experiments, and explored school and neighborhood gardens to learn more about their local bee population. "Because my students engaged in real-life problem solving based on their own scientific observations and data, their scientific investigation skills improved," said Panik. "I was especially pleased to see them asking questions, designing their own investigations and solutions, and using their data about the alarmingly low pollinator population to develop reasonable conclusions and possible solutions." The fifth graders' work culminated in "Bee Information Night" for the school and community, an informational presentation and fund-raiser to help them kick off their service-learning activities.

reinforces the need to avoid bias in scientific investigations—which is especially important if the environmental issue under study is controversial.

When students are learning science by studying environmental conditions in their community, region, state, or nation, they get the chance to apply fresh approaches to problem solving—they no longer sit passively just listening and taking notes. As they have more opportunities to become engaged and involved in real-world problems, they have the chance to refine their scientific observation skills as well as their ability to collect and analyze data and formulate rational conclusions. The effectiveness of EBE programs in developing students' scientific skills is demonstrated in numerous schools across the country in which students are responsible for gathering data, such as water quality samples, for local and state environmental agencies.

The gains in science achievement reported for students in schools implementing EBE programs are usually very substantial. The examples below are representative of the results from some of these research studies:

- Ninety percent of grade-level comparisons demonstrated higher scores on PACT tests by students in the EIC Model programs ($n = 10$ pairs of classrooms in ten South Carolina schools).[10]
- Average scores on the Colorado Student Assessment Program (CSAP) science assessment were 6 percent higher for students in schools implementing an EBE program over students in traditional programs within a four-county comparison group of schools ($n = 317$ students in program compared to weighted-average assessment scores from four feeder school districts).[11]
- In 64 percent of standardized test scores in science, students in the EBE programs performed better than students in traditional programs ($n = 11$ grade-level student assessment comparisons in eight pairs of California schools).[12]

History/Social Science

In traditional history and social science classrooms, students are often limited to textbooks and other paper or electronic documents, but with EBE programs, by examining local and regional environmental topics, they can more effectively learn about and apply their knowledge to current and historical events, laws and legislative processes, ancient and modern civilizations, economics, politics, and geography.

A Study in Land Use over Time

Third-grade teachers at Weaver-Odom Elementary in Aldine, Texas, used a study of land use in their community to teach two Texas geography-focused social studies standards. Students investigated how land use has changed the natural environment of their community over time. As a starting point, students learned the cardinal directions by drawing a map comprising their campus buildings, sidewalks, playgrounds, gardens, and nearby roads. They then examined both historical and present-day local maps and aerial photos and noted changes in the distribution of habitation. Teachers assessed student maps using more traditional tests to assess their knowledge of geographical vocabulary and map components.

In one notable example of EBE, the students and teachers at Armuchee Elementary School in Rome, Georgia, brought history/social science to life after a field trip to the Arrowhead Environmental Education Center. Kim Kilgore, one of the center's educators, spoke to the fourth-grade students about the importance of the green tree frog to Georgia's ecosystems. It was as a result of that conversation that the students and their teachers, Marilyn McLean and Ruth Pinson, started their quest to get this small frog some legal protection. The students began research and the teachers invited speakers to describe the process of how a bill becomes a law. The students concluded that they should ask Georgia's legislature to name the green tree frog the state's official amphibian. They lobbied locally and traveled to the capital hoping to garner support. McLean says, "They would walk up and talk to anyone who was interested." Despite all their hard work, they were not initially successful. It took more time and more students to get the state Senate's approval, ultimately a unanimous 52–0 vote. "In retrospect, maybe losing the first two years was a good thing," McLean reflected, "Our students learned the ins and outs of the political process."

Through EBE programs, students in history/social science begin to perceive the connections between history, politics, economics, and cultures. Comprehending these connections engages students more deeply in their studies and helps them see the significance of history/social science in their lives. When students are more engaged and have a better understanding of the connections between human social systems and the environment, they become more interested in and capable of asking their own questions about their community, their environment, and the world around them—ultimately making them better prepared to be involved and active members of our democratic society.

Authentic lessons such as those described in this chapter have been shown to do a better job of helping students retain history/social science content long after their examinations. Sue Fogel, now a retired seventh-grade social studies teacher at the middle school in Chariton, Iowa, put it this way, "The fact that students could still apply what they learned earlier in the year tells me that they've learned it well. I can see that they have brought information from their fieldwork into the classroom. These are things the kids are never going to forget . . . these big projects we do around Chariton, those are the ones they're going to remember down the line." Fogel attributed the students' improved retention to their EBE program and instructional activities like their river unit. Through this unit, students had the opportunity to study local geography, make maps of rivers and roads, explore where and how fur traders lived and worked, and to discuss how all of this connected with historical and present-day modes of transportation—all tied together with their scientific studies about issues of water quality and supply in their local rivers.

As students have the opportunity to use their history/social science knowledge and skills to resolve local environmental challenges, they begin to develop into catalysts for community improvement. Equally important, when they get a chance to work on projects like planning and developing a pocket park in Glenwood Springs, Colorado, they discover that they can make a difference. "I call what we're doing applied government," commented Guy Brikell, a social studies teacher at Glenwood Springs High School. "We're applying learning and giving the kids an opportunity not only to read a textbook about something, but to experience it. There's no other way we can replicate a lesson like that. It's not necessarily a pretty lesson, but it's a real lesson about how government works and communities operate."

Data are somewhat limited regarding students' gains in history/social science achievement for schools implementing these programs. The small number of relevant studies may be an artifact of federal No Child Left Behind legislation, which does not require student assessment of history/social science learning as part of states' analysis of Annual Yearly Progress (AYP). Below are two examples of relevant research studies:

- There was an 11 percent increase in the number of students performing at the satisfactory level on LEAP 21 in Louisiana's East Feliciana School District over a three-year period after implementing a place-based program.[13]
- Seventy percent of grade-level comparisons demonstrated higher scores on PACT tests by students in the EIC Model programs ($n = 10$ pairs of classrooms in ten South Carolina schools).[14]

STUDENT ENGAGEMENT AND BEHAVIOR

Students in EBE programs develop a passion for learning that they may never have had before, which leads to increased engagement and greater motivation to succeed. Not surprisingly, when students are excited and engaged in their learning, they begin to take more personal responsibility for their schoolwork. This pays off in many ways, including better attendance, less tardiness, and fewer classroom discipline problems. As one fifth grader in Ohio put it: "Even though I don't like school very much, I look forward to coming more than I used to. It used to be just sitting looking at books. It's better when you do stuff that you can actually see and not just read about."

In addition to enhancing learning, providing opportunities for students like the sixth graders at Huntingdon Area Middle School in Huntingdon, Pennsylvania, to become actively engaged rather than bored with a traditional lesson pays off in terms of student self-discipline. As a seventh grader in another Pennsylvania school said: "I seem to like doing experiments better than the sit-down, shut-up-and-write tests. I think I am a hands-on person." The vice principal at Huntingdon discovered this same thing when he compared students in the EBE program with students in traditional classrooms—students in what they called "STREAMS" had 97 percent fewer referrals to the principal's office than students in the traditional program.

The town of Red Oak, Iowa, may be in a rural part of that farm state, but it is not isolated from the environmental concerns found in large urban areas. When the students at Red Oak Middle School learned that a nearby U.S. Environmental

Protection Agency Superfund site was believed to be at least partially responsible for the town's high cancer rates, they became concerned. They became active in investigating the situation when they realized that this serious environmental issue was of great personal importance to them, their families, and their community.

Barbara Sims, then a Red Oak teacher and now the retired principal, built upon the students' interest by framing their science lessons around what interested them. "A lot of times, we do not give kids this age enough credit. I began by asking them what they wanted to know," said Sims. The students' engagement and enthusiasm was apparent, according to Sims, when they "began exploring where they could get more information . . . They would get on our speaker phone with lists of questions for different community resources." Implementing the EIC Model at the school was so important to her that, after she became principal, she worked to have it used throughout the school: "I would like to have my entire staff experience the camaraderie that we experienced and the revelation that this will engage students and help with retention."

Students in EBE programs like the one at J. R. Briggs Elementary School in Ashburnham, Massachusetts, are so excited about the work they are doing that their behaviors change. "Many students that have difficulty in the classroom bloom when they are outdoors doing hands-on activities," says Mary Gagnon, one of the school's third-grade teachers. Briggs teachers frequently give students the chance to do everything from sensory walks and poetry writing to standards-based research projects and reports on sites like the nature trail behind the school, a nearby vernal pool, and Dragonfly Pond. Students do not just visit the pool and pond one or two times; they get the chance to explore and study the sites many times throughout the year, from kindergarten through fifth grade. Ongoing opportunities like this increase engagement and give students a chance to build a sense of caring that can grow into a commitment to environmental stewardship.

During their investigations, the students at Briggs realized that a housing development and new roads were being built near the privately owned Dragonfly Pond. Concerned about the health of the pond and the animals that depended on it, they decided to write letters to the landowner, asking him to protect the pond by limiting the amount of construction near the pond and to give them permission to study the area. Their enthusiasm and arguments paid off. These fifth-grade students were so persuasive that the landowner changed his plans so that all Briggs students could continue to visit and study. He even added wheelchair access for students with limited physical abilities.

Few direct measures of student engagement and behavior, such as surveys of students, have been collected in schools implementing EBE programs. However, Scott Heydt, one of the fourth-grade teachers at Seven Generations Charter School in Emmaus, Pennsylvania, developed an instrument for just this purpose. Teachers there recently conducted a schoolwide study that provides student survey data that support what teachers and students report during interviews. In the survey, 76 percent of the K–7 students reported that they are more excited and/or productive when they are involved in their EIC Model community-based investigations.

Although there are limited data directly related to engagement and behavior, there are other data that can be used as proxies for this information. Changes in classroom behavior, for example, can be inferred in schools that record the number of disciplinary actions observed by teachers and administrators. Similarly, records of attendance and tardiness, and achievement motivation can be used as a proxy for assessing student engagement in school activities. The examples below represent some of the results currently available from studies of EBE programs:

- Classroom behavior—disciplinary actions
  - Middle school students in South Carolina's EIC Model program were 78 percent less likely to be the subjects of in-school disciplinary actions than their peers in traditional school programs (*n* = six South Carolina schools).[15]
  - There was a 5.7 percent lower insubordination referral rate for students in the EIC Model program compared to the tenth- through twelfth-grade students in a traditional Florida high school program.[16]

- Attendance, tardiness, and motivation
  - Seventy-seven percent of assessments of student attendance indicated that students in the EBE programs performed better than students in traditional programs (*n* = 22 grade-level comparisons in eight pairs of California schools).[17]
  - There was a 29 percent lower unauthorized absence rate for students in the EIC Model program compared to the tenth- through twelfth-grade students in a traditional Florida high school program.[18]

- There was a 10.8 percent lower tardiness rate for students in the EIC Model program compared to the tenth- through twelfth-grade students in a traditional Florida high school program.[19]
- Statistically significant higher scores were achieved on an assessment of "students' achievement motivation" for both ninth- and twelfth-grade students in EIC Model programs ($n = 400$ students in eleven Florida schools).[20]

WIDE-RANGING BENEFITS

Gains in academic achievement, increases in student engagement, and improvements in behavior are not the only benefits of EBE programs. Environment-based learning activities typically give students an opportunity to develop and apply higher-level thinking and problem-solving skills. As they explore how they and their communities might address an environmental challenge, students develop the kind of communication and leadership skills they need to work with community government. Ultimately, their environmental work can give them wide-ranging opportunities to learn from and work with scientists, historians, local businesses, professionals in the media, government agencies, community members, and leaders—allowing them to observe up close jobs and careers they might otherwise never see.

A good example of this process and the benefits that accrue occurred in an urban school in Massachusetts. In 2003, eighth-grade students at the New Bedford Global Learning Charter School in New Bedford, Massachusetts, became concerned when they found out that swimming and fishing in parts of Buzzards Bay were off-limits and unsafe due to the discharge of untreated waste from an aging sewage treatment plant. With the guidance of Christopher Jones, their history and law teacher, they undertook a study of how their community had been affected by the untreated sewage and how they could benefit from the construction of a new sewage treatment plant.

The resolution of this problem had been plagued for years by funding and legal issues, and students began digging into the environmental protection requirements that cities and towns have to follow under the federal Clean Water Act. They enlisted the help of a local attorney and former city councilor to help them explore how the law affected New Bedford. They met with a local judge, who explained legal aspects of environmental compliance and how the legislative and judicial systems interact with the executive branch of local government as well as with the U.S. Environmental Protection Agency. The students heard the specifics of the environmental concerns from the Coalition for Buzzards Bay, which had been collecting data about the bay's water quality for years. They learned that the existing primary sewage treatment plant had been releasing fecal coliform bacteria at unhealthy levels, making swimming and fishing in some parts of the bay unsafe and off-limits. They then compared these data to data collected from the secondary sewage treatment plant and discovered that the new system was already improving water quality in Buzzards Bay as well as in the local watershed.

Deciding to gather their own data, they began taking water samples from the upper watershed, the Acushnet River, and the harbor to measure dissolved oxygen, temperature, and turbidity, and entered their data on the government-sponsored GLOBE website. They followed up by visiting a local wastewater treatment plant to learn about how wastewater is treated and brought to legally acceptable levels.

Proud and excited by their work, the students shared their results with the school at an end-of-year presentation on learning, and culminated their project with a community education campaign that included the creation and distribution of informational pamphlets about water usage trends and ways to conserve water.

A FINAL BENEFIT

School and district administrators find that the benefits from EBE programs are great enough that they are willing to provide their teachers with the time to develop innovative programs and to identify, design, and participate in a wide range of professional development opportunities. Rodney Stewart, principal at Armuchee Elementary School in Rome, Georgia, and his predecessor felt so strongly about this that over a four-year period beginning in 2002, they had all of their teachers trained in developing the school's EIC Model program. "The way we taught yesterday is not good enough today," he says. "We know the price up front is higher for incorporating best teaching practices from the EIC Model. We are also aware the educational rewards are far greater. Once you see the kids' enthusiasm, work becomes so much fun." In the next chapter, we will explore three examples that illustrate how a school, district, and state created and implemented EBE programs designed to meet their specific educational needs.

CHAPTER FOUR

Implementing EBE in a School, District, or State

Environment-based education (EBE) programs have been established for many different reasons: to start or restructure new public and charter schools; to create a districtwide school choice program; and in one state, to develop an innovative statewide curriculum. Notwithstanding their different scales, scopes, and unique educational challenges, three of these programs—Arabia Mountain High School, Desert Sands Unified School District, and California's Education and the Environment Initiative—are good examples of how this innovative approach to learning can be successfully implemented.

In chapters 5–9, we will go into further detail about how to design and implement an EBE program at a classroom, school, district and statewide level—a project that involves similar steps no matter what the scale or the particulars of the local context.

ARABIA MOUNTAIN HIGH SCHOOL

When DeKalb County's Board of Education and the school district office were presented with a unique opportunity to establish a new high school on land within the Arabia Mountain National Heritage Area, they decided to take advantage of their good fortune by building green, and they became Georgia's first school to receive Leadership in Energy and Environmental Design (LEED) certification.[1]

Long before construction began, however, visionaries in DeKalb County's school system decided to go much further than simply building an environmentally healthy facility. They wanted to capitalize on the opportunities that would be presented by locating the school adjacent to the Davidson-Arabia Nature Preserve and Panola Mountain State Park. They decided the school would be an "environmental school."

To launch the project, Horace Dunston, the county's assistant school district superintendent for the area, became the district's liaison to both the state park and the local community. He met repeatedly with these stakeholder groups prior to construction to get input on the design of the school and to learn what limitations on construction and eventual student use would result from locating the facility within the Arabia Mountain National Heritage Area. These discussions were the starting point for the development of strong working relationships with and ongoing involvement by these groups on a variety of joint project activities, such as funding for studies of the natural systems of the Heritage Area and Arabia Mountain.

Wanda Gilliard, the district's associate superintendent for curriculum and instruction, knew that creating a school with an environmental focus meant finding an educational strategy that would allow the development of a fully integrated curriculum. In giving the task to Faatimah Muhammad, the district's science coordinator, she knew she was giving the job to someone who lived in the area and was familiar with it. Muhammad began searching for an appropriate EBE strategy and soon came across the Environment as an Integrating Context for learning (EIC) Model and Georgia's EIC Model Demonstration School Network. After speaking with some of the teachers in the network, looking at literature reviews, and examining the State Education and Environment Roundtable's (SEER) evidence about the EIC Model, she took her recommendation to Dr. Gilliard and the district office. That is when they decided to enlist SEER's assistance and implement the EIC Model as the framework for the curriculum at the new high school. With only nine months remaining before the school would open its doors to students, the process went into high gear.

With community support and instructional design well in hand, the DeKalb County School District proceeded with their implementation plan. The construction of the school began in 2007 through the district's normal funding channels. With the size of the school and the inclusion of all of the features required to

EIC Model Demonstration School Networks

In 2000, SEER began developing a series of state-level EIC Model Demonstration School Networks in collaboration with its member state departments of education and two nonmember states. Altogether, SEER was able to establish demonstration school networks involving 135 schools in thirteen states, including: California, Florida, Georgia, Idaho, Iowa, Maryland, Massachusetts, Minnesota, Montana, New Jersey, South Carolina, Texas, and Wyoming. (Note: In 2013, after more than thirteen years, not all schools in the original networks are still involved.)

SEER and its state partners gave teaching teams representing all participating schools a professional development institute that provided four days' instruction about the model, and assistance in designing their first EIC Model instructional units. The institutes were followed by team planning sessions at the schools and site visits and technical support from state representatives and SEER staff members during their first two to three years of implementation. This development process was also supported through formative and summative evaluations based on SEER's EIC Model evaluation instruments (discussed further in chapter 9).

The networks were designed to demonstrate the efficacy of the EIC Model to educators in the participating states. One important function of the networks was to provide opportunities for other schools to visit and observe the programs; therefore the process of selecting schools sought to include a widespread geographic area in each state. In Georgia, for example, the Department of Education invited all public schools in the state to apply for membership in the network. Ultimately, the State Board of Education approved ten schools based on their commitment, potential for success, and demographic diversity: Armuchee Elementary, Floyd County; Midway Elementary, Baldwin County; Minor Elementary, Gwinnett County; Saint Simons Elementary, Glynn County; Shakerag Elementary, Fulton County; Arnold Middle, Muscogee County; Gainesville Middle, Hall County; Henderson Middle, DeKalb County; Columbia High, DeKalb County; and Stewart-Quitman High, Stewart County.

The configuration of EIC Model networks varies from state to state, as they are characterized by a variety of funding sources, implementation strategies, and support mechanisms. The development of each state-based network has provided valuable information about:

- structuring the network of demonstration schools to best support the administrators and teachers;
- designing the professional development workshops to meet the varied needs of the participating schools;
- successfully facilitating the school change process at each school site;

(continues)

- forming collaborative instructional relationships among administrators, teachers, parents, students, and community members;
- most effectively documenting the EIC Model implementation process; and
- reporting academic and programmatic successes to appropriate stakeholders.

The importance of demonstration school networks in the development of "new" EIC Model schools has been seen in several recent cases. For example, representatives for the new EIC Model school, Arabia Mountain High School had observed schools in the Georgia network, and most recently, a group from a new EIC Model school in Colorado had visited Seven Generations Charter School in Emmaus, Pennsylvania.

achieve a LEED-certified building, the total construction costs reached about $50 million. In addition to construction costs, the district provided the majority of the funding for the professional development and technical support that the teachers would need to develop their EIC Model program. These resources came from their professional development budget. Several nongovernmental organizations, philanthropic groups, and local businesses also provided medium to small grants that paid for follow-up professional development and student research projects. Over the years, numerous community partners have continued to provide a wide range of services on an in-kind basis.

Hiring began in earnest. Dr. Angela Pringle, at the time a principal at another DeKalb County high school, was chosen to lead Arabia Mountain High School. "We made sure that she got indoctrinated in the details of the school, the community, LEED certification, and the curriculum framework we had selected for the school, the design of the school, and how the students would be selected to come to the school," says Dr. Gilliard. "We gave her an overview so she could begin educating the parents and students who would be interested."

Almost immediately, Dr. Pringle began choosing her department chairpersons. Fortunately, she found highly experienced educators who were willing to take on the challenge of starting a new school. "We wanted to build the infrastructure from day one, so all of the department chairs and the central office coordinators were the first ones trained in the design of curriculum based on the EIC Model," explains Dr. Gilliard. "[This way] every one of the key department chairs who would have meetings with the community, with the parents, and with the students would all be saying the same thing. They needed to know the instructional design and understand that the EIC Model was the priority."

School leaders and district staff members continued to invest time to keep key stakeholders like policy makers, community leaders, park representatives, and parents informed and up-to-date. Dr. Gilliard described their step-by-step process by explaining that they "did a special overview for members of the Board of Education so that they would understand about the dynamics of the school, LEED certification, selection of the EIC Model as the framework for the curriculum, the student selection process, and all of the key components. The school's growing faculty and district staff were all prepared so that when the community asked them questions, they would have a 'script' and give the community the same information."

One of the most important elements of this process was achieving a shared understanding of the high school's overarching goals. In early 2008, the school's administrators and coordinators met with senior district staff members to develop a mission statement that, to this day, describes the school's purpose:

> The mission for Arabia Mountain High School is to engage students in active learning and service through collaborative instruction using the school and community as context for developing understanding of human and environmental interactions while preparing students for involved citizenship in a changing world.

With the school's leadership team and mission statement in place, and with the full backing of the DeKalb County Board of Education, Dr. Pringle and her department chairs began hiring the teaching teams. She took care to be open and clear during this hiring process. Dr. Pringle described her ideal candidates by saying, "We are seeking teachers with passion and ability who are willing to work in a highly collaborative environment." From the very beginning, she told all potential applicants that a major part of the job involved developing a curriculum that helps students succeed with Georgia Performance Standards using the EIC Model as the curriculum framework. Dr. Pringle also let them know that they would be required to participate in a four-day EIC Model professional development institute during the summer.

The district office gave Dr. Pringle a great deal of flexibility in hiring, allowing her to bring in teachers from other schools in the district as well as to conduct open searches. Although her highest priority was to find teachers who were already certified, she let it be known she would also consider talented scientists or engineers willing to become certified teachers.

That summer, SEER staff members provided an EIC Model institute for all of the teachers, counselors, and administrators at the school. Community partners like the Georgia Department of Natural Resources, Georgia Power, and the Flat Rock Archives museum, who planned to work closely with the school, were also invited to participate so that they, too, could learn about the curriculum that the school was starting to develop.

In September 2009, with the vast majority of its sixty ninth-, tenth-, and eleventh-grade teachers, counselors, administrators, and community partners in place and trained to develop their EIC-based curriculum, Arabia Mountain opened its doors to their first group of over one thousand ninth-, tenth-, and eleventh-grade students from across the county. During the following two years, the school grew to include twelfth grade, and hired and trained thirty new teachers.

Hiring and preparing new faculty members was by far the school's greatest challenge. Despite all of Dr. Pringle's efforts to ensure that applicants understood "what they were getting into," many teachers did not clearly understand the school's mission or instructional strategies, and their classroom instruction began to look the same as it would have in a traditional school. Some of them, for example, were not comfortable taking their students out into the community to develop their environmental investigations, while others had difficulty working as grade-level teams to integrate instruction in multiple disciplines. The school's administrators worked to resolve this situation in two ways. Dr. Pringle appointed one of her assistant principals, Tim Wells, to coordinate all aspects of the EIC Model program implementation. Wells, in turn, worked to ensure the faculty received ongoing technical support by means of both video conferencing and site visits by SEER's professional development team. This gave teachers the opportunity to ask how other schools had resolved some of the issues they were facing and to get input on the design of their EIC Model instructional units, especially the design of community-based investigations and integrated interdisciplinary instruction, both of which they were finding challenging.

Despite this difficult start, today the school is doing well. In 2011, the school's students, with a demographic composition comparable to the state averages (over

50 percent free/reduced-price lunch), beat the state averages on Georgia's Criterion Referenced Competency Tests (CRCT) in English language arts, biology, economics, and U.S. history, just as they had in 2010.

Dr. Gilliard reports that: "The school has done very, very, very well because many of the leaders who were originally trained are still there . . . so even though the principal has changed, the teacher leaders have stepped up their leadership roles in their individual content areas. They are the ones making sure that there is a high level of curriculum instruction, high level of rigor, and the high level of program that we want there."

All of the preparation by the teachers and administrators paid off when students began to create exciting environmental investigations in the community. As part of an eleventh-grade energy unit, for example, Arabia Mountain students identified a desire to reduce energy consumption at the school and among local businesses. They selected businesses that interested them, like a dentist's office, hair salon, automobile repair shop, and medical clinic, and conducted energy and environmental audits of their operations. Students conducted cost analyses and provided the businesses with reports, including recommendations to make the business more "environmentally friendly." One of the students involved in the study at the hair salon even received a grant to help the owners implement some of her recommendations for energy conservation and reduce the use of toxic chemicals used in various hair treatments.

In 2012, Arabia Mountain High School was recognized with a U.S. Department of Education Green Ribbon Schools award. "This goes to show that schools in Georgia are striving to be more energy efficient, developing healthy living habits and providing environmental education," said State School Superintendent Dr. John Barge.

Educators also see great value in terms of the transition to newly adopted state standards. Tim Wells, an assistant principal at Arabia Mountain, explains that "with Georgia's growing commitment to the Common Core State Standards, and the increased emphasis on students using their knowledge and skills in real-world situations, EIC has become an important way to demonstrate that they learned the content and skills and were able to use them." He also mentioned that "EIC has put our school [very] far ahead with the integration of Common Core State Standards, because we are already using those components."

DESERT SANDS UNIFIED SCHOOL DISTRICT

All too often, mission statements are written, posted on the wall, and quickly for-gotten. This has not been the case in Southern California's Coachella Valley, where a district of twenty-seven schools lives its mission statement every day:

> The mission of the Desert Sands Unified School District is to ensure that every stu-dent develops the knowledge, skills, and motivation to succeed as a productive, ethi-cal, global citizen by assuring equal access to student-centered learning provided by caring, committed, collaborative staff working in partnership with our families and diverse communities.

In 2000, Dr. Doris Wilson, at that time the superintendent of Desert Sands Uni-fied School District, took the mission to heart, envisioning educational programs that would provide high-quality learning opportunities for all students in her demographically diverse district, where 58 percent are eligible for free/reduced-price lunch, and 27 percent of the students are English language learners. Dr. Wilson and her colleagues, with strong support from the school board, began the process of creating a series of magnet and choice schools across the district with the goal of improving learning outcomes for their students.

The Desert Sands team began by restructuring three traditional schools into one International Baccalaureate magnet program located in the town of Indio in the eastern Coachella Valley. Their initial target was to serve the high-achieving students from Indio—the most economically challenged of the six communities that make up the district. With demand for this new educational opportunity out-stripping availability, another elementary school and a middle school were also transformed into International Baccalaureate schools.

In 2001, the Desert Sands team came across an opportunity for federal funding under the Voluntary Public School Choice Program. This time, when they surveyed parents across the district to gauge their thoughts about different educational alter-natives, they found that 85 percent wanted their children involved in programs that emphasized science and the environment.

With this in mind, Desert Sands administrators (backed by the Desert Sands School Board) focused their grant proposal for school choice funding on a group of six schools (one middle school and high school along with their four feeder elementary schools). Since their proposed environment-based education program would be serving students across the district, they took care to account for transportation costs in their grant request. Meanwhile, they kept parent and community representatives actively involved in the program design process by inviting them to regular meetings, where they discussed program goals, priorities, and ideas for curriculum design.

Superintendent Wilson and Darlene Dolan, assistant superintendent for curriculum, instruction, and assessment, designed the program with a specific vision in mind: "To provide students with meaningful experiences that broaden their horizons and connect the knowledge they gain through the study of 'environmental science' to their choices throughout their lives." The district's commitment to the program was backed by parent interest and by pledges of participation and support from diverse organizations, including local universities, museums, environmental groups, professional associations, and cities, in addition to regional and federal agencies. Wilson and Dolan's work paid off when the U.S. Department of Education's Office of Innovation and Improvement selected the program for a five-year $9 million grant.

As luck would have it, this occurred in 2001, when California State Senator Tom Torlakson's bill establishing the School Diversion and Environmental Education Law was enacted with funding from the California Integrated Waste Management Board. This law focused on getting school districts to integrate school-site waste management and recycling into their academic program. Seven districts were selected to become "Environmental Ambassadors" and receive two-year grants of up to $90,000. The state also allocated resources so the districts would receive two years of professional development and technical support as they designed their programs.

Assistant Superintendent Darlene Dolan's plan, and the district's commitment, was to pool their federally funded "choice" program monies with the state's funding to implement the EIC Model, with ongoing professional development and technical support from SEER and the waste management board's staff. After a series of conversations, site visits, and the completion of an extensive application, Desert Sands was selected to become one of California's "Environmental Ambassadors." With the state and federal funding in place, Desert Sands embarked on its multiyear journey to implement what they called the "Choices" program. In the spring of 2003, the district's plan took final shape and was put into action.

In August 2003, eighty teachers and administrators came together in grade-level or interdisciplinary teams representing each of the six schools. Over a five-day period, they actively participated in an EIC Model professional development institute, where they were introduced to all the instructional components of the model as well as to service learning, collaborative instruction, student assessment methods, and program evaluation. Then, working in teams, teachers and administrators designed standards-based curriculum units that would engage their students in the classroom, and that included learning experiences in the community through various field studies and service-learning projects. At the high school level, teachers also built in a variety of school-to-career opportunities that partnered students with a wide array of local and regional agencies offering experiences including internships, research, and job shadowing. The high school also added an Advanced Placement environmental science course.

One thing became abundantly clear from the beginning of the professional development institute: participants needed to define in their own terms what the district meant when it wrote about using the environment to broaden students' experiences so that they could apply their learning and make good choices throughout their lives. As the driving focus of the combined Choices and Environmental Ambassador programs, this became the most important, and difficult, conversation they had all week. There was a great deal of debate about the scale of the environmental issues students should investigate. Some teachers were more comfortable with relatively simple environmental topics, like gardening, while others wanted students to look at the big picture of natural and human social systems' interactions. Another major discussion revolved around whether all six schools would share the same set of "organizing questions" on which they could focus all of their students' community-based investigations.

Eventually, the teachers, school administrators, and district staff members decided that a systems-thinking approach would be best, and that their organizing questions should reach across the district. As difficult as this discussion may have been, it clarified the program's overarching goal and guided all of their instructional planning. Desert Sands educators reached a consensus that the

Desert Sands Unified School District's Organizing Questions

- What are the natural (physical and biological) and human social (individual, cultural, political, technological, governmental, and economic) systems in the Coachella Valley area, and how do they interact?
- What resource-use and management issues result from the interactions of the natural and human social systems of the Coachella Valley area?
- What past, present, and future individual, community, and political decisions affect the natural and human social systems of the Coachella Valley?
- What natural and human social systems in the Coachella Valley area contribute to the advancement of the fine arts, to the development of recreational opportunities, and to the promotion of esthetic values?

organizing questions should be designed to connect the schools' environment-based studies with both the natural systems of the surrounding desert and local decision-making processes.

But even as the program continued to develop, not all school site administrators were equally committed to implementing it, and many teachers complained that they were not fully informed about the goals of the grant programs or about the specific instructional strategies that were going to be implemented, or that they were going to be responsible for developing their own integrated and articulated K–12 curriculum. Although this subject had been addressed during the professional development institute, discussions continued during planning sessions over the first several months of the program. Many participating teachers also felt they needed greater technical knowledge about the environment before they could fully implement the program, including knowledge about local plants and animals, desert ecosystems, local environmental concerns, or even how to find good places they could take their students in order to conduct studies. Strong support from local environmental agencies and organizations like The Living Desert, Coachella Valley Wild Bird Center, and PEAK, The Energy Coalition helped to educate the teachers about all of these local topics.

In spite of these hurdles, Desert Sands' Choices program got off to a strong start. The key to this success was the convergence of the educational and policy interests of a diverse group of stakeholders:

- Desert Sands' school board, which had approved the federal and state funding requests because they viewed the program as a means of achieving the district's overarching mission of helping students become proficient in academic standards while preparing them to be decision makers.
- The U.S. Department of Education's five-year grant intended to create models of state and district programs that offered choice among different public schools.
- And the California Integrated Waste Management Board's grant to determine if an education-centered policy to increase school and district waste management could achieve the same goals as a regulatory policy approach.

Over the decade since the Desert Sands environmental program was initiated, it has evolved at each of the six schools: some have expanded their programs, while others have taken on a different focus. The Palm Desert Middle School, for example, was transformed into a charter school, where the mission includes academic excellence, environmental stewardship, and social accountability as global citizens. Its math department is engaged in interdisciplinary studies that give students the opportunity to improve their math skills through real applications of math concepts. Field studies, for example, include having sixth-grade students create their own environmental service-learning projects in cooperation with the Joshua Tree National Park, the Coachella Valley Preserve, Big Morongo Canyon Preserve, and several other local organizations. For the past three years, 2010–2012, the school's students have attained the highest Adequate Yearly Progress (AYP) scores of all fourteen middle schools located in the Coachella Valley. Equally important, three of the four participating elementary schools consistently place among the highest-scoring schools in the state's accountability system.

CALIFORNIA'S EDUCATION AND THE ENVIRONMENT INITIATIVE

Heal the Bay, a California environmental group established in 1985, started out dedicated to protecting and educating the public about the coastal and marine life in and around Southern California's Santa Monica Bay. Over the years, the group's interest in bringing content related to coastal and marine environments to children kept growing, and they wanted to expand their efforts into schools. By 2003, as their leaders began examining the state's educational standards, they realized California's standards contained little academic content about marine life and other environmental concerns.

Leslie Mintz (Tamminen), Heal the Bay's legislative director and advocate to the state of California, brought this concern to Assemblymember Fran Pavley—a former classroom teacher and longtime supporter of environmental protection. Their discussions eventually resulted in a draft bill that called on the State Board of Education and the Department of Education to formally include ocean- and environment-related content in the state standards.

California's education officials reacted negatively to the approach that Heal the Bay proposed, since they had only relatively recently completed the process of creating California's academic standards and begun adopting conforming instructional materials. Over the following months, Pavley and Mintz met time and time again with representatives of California's education and environmental agencies. Ultimately, these discussions led Assemblymember Pavley to author Assembly Bill 1548, a compromise among all of the stakeholders. Rather than requiring additional education standards, the compromise legislation called for the development of "education principles for the environment" that would be taught by integrating the state's existing science and history/social science standards into new curriculum units for use by schools. These principles would be developed by the state's Environmental Protection Agency and Integrated Waste Management Board—the entities that would then be responsible for developing a model curriculum and working to get it implemented by California's 1,057 school districts. Passed by California's Assembly and Senate in the fall of 2003, it was signed into law in October by Governor Gray Davis. This law would eventually be called the "Education and the Environment Initiative" (EEI).

However, the discussions, and in some cases arguments, did not stop with the signing of the bill. In spite of the many conversations and compromises, the six agencies identified in the legislation as participants in this endeavor (State Board of Education, Department of Education, Governor's Secretary for Education, California's Environmental Protection Agency, Integrated Waste Management Board, and Natural Resources Agency) had significantly different views of the details of what implementation should look like. As is their legal responsibility, the education agencies were primarily focused on implementation of California's standards, using state-adopted instructional materials. At the same time, on the environment and natural resources side of the equation, the representatives remained principally concerned about the environmental content that students would receive under the EEI.

This situation was made worse by the fact that the legislature had been told that the work to be done under the law, including the development of the model K–12 curriculum, could be done at no additional cost to the state. Fortunately, that situation did not last long, because of the strong commitment to the environment of newly elected Governor Arnold Schwarzenegger and Secretary of the Environmental Protection Agency Terry Tamminen, as well as the waste management board chair, Linda Moulton Patterson, and her principal adviser, Bonnie Bruce. Working together, this high-level group of supporters committed themselves to seeking funding in the next budget cycle. As a result, California's next state budget included funding for the EEI through the Integrated Waste Management Board, and was supported by the governor's office and passed by the legislature.

In February 2004, Integrated Waste Management Board's newly established Office of Education and the Environment engaged SEER and The Acorn Group as the coconsultants for the EEI. With the full team in place, the work of implementing the EEI began.

An early objective was broadening the constituency for the EEI. Environmental Protection Agency staff members and EEI consultants did this by arranging a series of meetings with representatives of the State Board of Education, the California Department of Education, and the governor's Secretary of Education. Several state

Costs of Implementing the EEI

It is difficult to provide an accurate estimate of the total costs of implementing the EEI. The early stages of development principally involved the time of the coconsultants. The phase of developing the EEI curriculum involved the consultants, numerous writers and editors, cartographers, photographers, and graphic design teams. This five-year process cost in excess of $10 million. The final phase, distribution of the curriculum and the associated professional development, has not been completed; however, early estimates of the cost range up to $30 million.

board and department of education staff members remained concerned, however, about how the "Environmental Principles and Concepts," previously called the education principles, would mesh with curriculum required by state standards in English language arts, math, science, and history/social science. In fact, a State Board of Education senior staff member, who a few years later became a board member, remarked, "I hate this law."

The conversations that took place during that first year were a vital part of efforts to build good working relationships among all the state agencies and the EEI's lead consultants. One of the most essential elements of that process involved demonstrating to the educational decision makers that the team working on the EEI (consultants and agency staff members) fully understood the importance of the state's educational priorities, including state standards, curriculum frameworks, student assessments, and State Board of Education policies. A second important component of this process involved building relationships with California's environmental organizations, business communities, and professional education organizations, like the California Science Teachers Association. The EEI team conducted frequent meetings with all of these stakeholders throughout the eight-year process of developing the state's Environmental Principles and Concepts and creating the EEI curriculum—this set of relationships proved fundamental to the ultimate success of the endeavor.

The first phase of implementation required developing mutually agreed upon goals for the environment-related education content, the Environmental Principles and Concepts—the framework for student learning that would ultimately guide the development of the curriculum. Over 100 environmental experts participated in this process as representatives of state and federal agencies, universities, industry, and not-for-profit environmental organizations. (See appendix: "Organizations That Participated in the EEI Technical Working Groups" at the end of this chapter.)

Using the initial input from this wide-ranging group of experts, the EEI team developed drafts that were subjected to multiple reviews, made available for comment on the Internet, and discussed at a series of focus-group sessions for the public across the state. Dr. Charles Munger Jr., then a member of the Curriculum Development and Supplemental Materials Commission (an advisory body to the State Board of Education on curriculum frameworks and instructional materials), also offered his scientific expertise in reviewing the draft Environmental Principles and Concepts—another critical element in building strong relations with the state's education decision makers. After this thorough vetting and review process, in accordance with the law, the Integrated Waste Management Board and the secretary of the California Environmental Protection Agency approved the final draft of five environmental principles and fifteen supporting concepts.

With this in hand, the EEI team began the process of designing the K–12 curriculum. One of the first steps was to survey 10,000 California teachers for their guidance. The survey requested input on: the design of the curriculum; the form in which teachers would prefer to receive the final materials; how teachers made use of adopted instructional materials; and even the types of professional development teachers thought would be most effective.

At about the same time, the EEI team established an Interagency Model Curriculum Planning Committee that brought together representatives of all six state agency partners. This committee was intended to serve two key functions: to provide input on the overall design of the curriculum, and to continue to strengthen the working relationships among all the key stakeholders.

Over the next several months, staff of SEER and The Acorn Group developed a comprehensive plan for developing the EEI curriculum. This plan provided hundreds of detailed learning objectives designed to effectively teach the state standards in science and history/social science in the context of the curriculum framework. This "Model Curriculum Plan" also included other major design elements, such as a scope and sequence, specifications for instructional units, strategies for student assessments, and plans for aligning the EEI curriculum with state-adopted instructional materials.[2]

The interagency committee reviewed and approved the proposed Model Curriculum Plan. Essential to obtaining this approval was inclusion of elements that met the needs of each of the key agency partners. The education agencies needed the curriculum to help students become proficient with standards-based content, make appropriate connections to adopted textbooks, and use appropriate student assessment strategies, including written assessments designed with the "look and feel" of the state Department of Education's standardized tests and performance-based assessments. The environmental agencies were focused on a plan that would help students understand the relevance of the Environmental Principles and Concepts to their decision making and daily lives. Figure 4.1 presents the workflow of major EEI activities and the key roles of the participating agencies, individuals, and committees.

FIGURE 4.1

Workflow of major EEI activities and key roles

Management and oversight responsibilities:
California Environmental Protection Agency (CalEPA)
California Integrated Waste Management Board (CIWMB)
Design and development of EEI plans and instructional materials:
EEI consultants (State Education and Environment Roundtable and The Acorn Group)

Workflow of major activities	*Roles of participants and committees*
Legislation passed and signed into law, 2004	• **Heal the Bay:** proposed legislation • **Assembly member Fran Pavley and Senator Tom Torlakson:** sponsored legislation • **Governor Gray Davis:** signed bill into law
Outreach was conducted to obtain input from stakeholders on implementation of the law	• **Education partnership and statewide focus groups:** provided general input on all aspects of the EEI • **CalEPA and CIWMB:** identified stakeholders, coordinated and managed efforts • **EEI consultants:** analyzed input
Environmental Principles and Concepts (EP&C) were drafted; reviewed by agencies, organizations and general public; finalized and approved	• **Technical working groups:** gave input on content of EP&C and reviewed drafts • **CalEPA and CIWMB:** reviewed and approved EP&C • **EEI consultants:** developed and revised EP&C
Educator Needs Assessment was undertaken	• **Classroom teachers:** 10,000 individuals were asked to complete survey • **CalEPA and CIWMB:** reviewed and oversaw survey process • **EEI consultants:** developed and analyzed survey
Model Curriculum Plan (MCP) developed and finalized	• **Interagency Model Curriculum Planning Committee:** gave guidance, reviewed development • **CalEPA and CIWMB:** reviewed and oversaw development • **EEI consultants:** developed and revised MCP, including learning objectives
Initial versions of the EEI curriculum were drafted based on the MCP	• **CalEPA and CIWMB:** oversaw drafting of initial versions • **EEI consultants:** managed curriculum development and implemented revisions
Drafts of each EEI curriculum unit underwent field-testing by up to five classrooms of teachers and students	• **CIWMB:** managed and analyzed field-testing

(continues)

Workflow of major activities	Roles of participants and committees
Drafts of each EEI curriculum unit were subjected to review by technical experts	• **CalEPA and CIWMB**: identified and managed experts • **EEI consultants**: analysis and revisions to drafts
Revised drafts of each EEI curriculum unit were produced based on field testing and expert comments	• **CalEPA and CIWMB**: oversaw revisions • **EEI consultants**: managed curriculum development and revisions
Drafts of all EEI curriculum units were subjected to in-depth review and analysis	• **State Department of Education**: managed and oversaw review and analysis of drafts • **Curriculum Development and Supplemental Materials Commission**: reviewed draft EEI curriculum by education and technical experts and gave recommendations to State Board of Education
Final drafts of all EEI curriculum units were submitted for review and approval by the State Board of Education	• **State Board of Education**: reviewed draft EEI curriculum and gave approval subject to changes recommended by the Curriculum Development and Supplemental Materials Commission
All EEI curriculum units were revised based on the resolution passed by the State Board of Education, which required final edits	• **State Department of Education**: oversaw revisions • **CalEPA and CIWMB**: managed and oversaw revisions • **EEI consultants**: implemented revisions
The approved EEI curriculum is being disseminated and professional development provided to educators across California	• **California's Department of Resources Recycling and Recovery** has been moved under CalEPA and is overseeing this work

Achieving agreement by the six state agencies gave legitimacy to the Model Curriculum Plan and meant that SEER and The Acorn Group could work with its group of sixty writers and editors using this preapproved framework to guide the development of all EEI-related instructional materials throughout the multiyear course of the effort. Creating this mutual understanding avoided many of the challenges that typically face statewide educational endeavors, such as midcourse changes to learning objectives and foundational instructional practices like hands-on learning. It also proved to be a vitally important part of every subsequent process, starting with the drafting of instructional content and concluding with review and approval of the eighty-five curricular units by the State Board of Education.

The time invested in building these relationships ultimately led to the approval of the EEI curriculum. That this effort paid off was clearly shown when the former staff member, who now, six years later, had become a member of the State Board of Education, said, "Since I was against this at the beginning, I feel I should be the one to make the motion to approve this curriculum"—this from the same individual who, seven years earlier as a senior staff member, had said, "I hate this law."

It was now time to roll the EEI curriculum out to teachers and students across California. This process began in earnest in early 2011 and continues even now, with the ultimate goal of getting the curriculum into the hands of every teacher in every one of California's 1,057 school districts.

COMMON CHALLENGES

Whether at the school, district, or state level, successfully promoting an environment-based education program involves many challenges—some common, others more dependent on the scale of the intended program. Involving all the decision makers and keeping them informed is a crucial element of the process.

It is important that, at the start, the stakeholders achieve clarity about the educational purposes of the initiative. This is necessary in order to engage many of the individuals and entities that will be involved in making the "go or no-go" decisions about the program. Equally important, the support and cooperation of many of these agencies and organizations may be required for the successful implementation of the program. If their concerns and perspectives are appropriately considered, they— like the senior state board of education staff person and future board member who was initially so firmly against California's EEI—may eventually be counted among the program's strongest supporters.

Potential program supporters and partners also need to have a clear understanding of the expected benefits of implementing a new EBE program. This has been SEER's experience when state agencies like Georgia's Board of Education agreed to support the development of an EIC Model Demonstration School Network; when the DeKalb County School District decided to implement the model at Arabia Mountain High School; and when Desert Sands Unified School District made their decision to implement EBE under their Voluntary Public School Choice grant. Comparable questions have always been asked by administrators of each school, school district, or state that has expressed interest in implementing SEER's program. The individuals involved in these decisions want and deserve evidence of potential benefits like those provided by the research described in chapter 3. This base of research makes them more comfortable with their decisions about implementing innovative instructional programs like EBE.

Another common challenge involves the need for programs to pay constant attention to both the perspectives they present and the language they use. Programs that seek to bring a political tone into the classroom generally receive little support from or credibility in the educational community. Very few schools, districts, or state education agencies will implement programs that promote an agenda or appear to be one-sided. If this aspect is not handled carefully, opponents to EBE programs can be expected to raise questions about biased materials they see as seeking to turn students into "wild-eyed environmentalists"—which may be an appropriate charge if the programs are not based on scientifically sound information that presents diverse perspectives. California's EEI, for example, was careful in its choices of writers, editors, and even words—all materials were reviewed several times to ensure that they were not presenting a biased point of view.

In addition to these wide-reaching common challenges, there may be issues specific to the level at which an EBE program is being considered: schoolwide, districtwide, or statewide.

Challenges for Schoolwide EBE Program Implementation

There is no set formula for getting an environment-based education program started in a school, school district, or state. Individual situations vary tremendously, and each requires its own strategy, partners, and timeline. Many of the challenges related to implementing an EBE program at the individual school level revolve around the seven primary concerns most often voiced by teachers and administrators.

Schools that intend to start a new EBE program need to invest the time to inform their school community about the program—its design, purpose, time frame, and how it will affect both students and teachers. If this is not done, new programs are likely to face challenges before or during implementation. In addition

Challenges to Implementing EBE Programs at the School Level

The primary concerns most often cited by teachers and administrators in relation to implementing an EBE program include:

- understanding the diverse perspectives of parents, administrators, and community members regarding EBE programming;
- confidence about having the authority, from district administrations or school boards, to implement a new program;
- insecurity about the level of support from school leadership or, for administrators, commitment from the district office;
- availability of necessary program coordination and implementation support during program development;
- inadequate "technical" knowledge about environmental content or the community where the school is located;
- limited knowledge or time to get community partners involved in implementation;
- ambiguity about what they are getting into, and a desire to have multiple examples of successful programs from which they can learn.

to informing the community about basics of the program, it is vital to seek input from and understand the thoughts and diverse perspectives of educators, administrators, parents, and other community members. Devoting time to listening to the community, as Horace Dunston, DeKalb County's assistant school district superintendent, did through community meetings, can be invaluable.

Basic to starting any new instructional program at a school is gaining the necessary authority and support from district administrators and school boards. This authority and support allows teachers and school administrators autonomy in, for example: applying for funding or joining a pilot program; adjusting class schedules; taking students outside of the school building or on field trips; bringing outside experts into the school; or developing partnerships with local businesses and community organizations.

Pennsylvania law, for example, requires that school charters be granted in the district where the school is going to be located. Even as the founders began planning the EBE curriculum for the Seven Generations Charter School (as detailed later in part two), the school had to be authorized by East Penn School District. The district had a very detailed application process that required the founding parents and community members to step outside of their normal range of experiences. Rejected the first time, the founders revised their application and resubmitted it to the district for approval. East Penn continues to be involved in high-level oversight of the school, as had been established in the original charter. In 2012, East Penn thoroughly reviewed the instructional program and student outcomes at Seven Generations and renewed the school's charter for an additional five years.

Teachers often report that they do not feel confident that they will have the necessary time, funding, administrative support, or other resources to implement a new program. Planning EBE programs can require a substantial investment of time for teachers, who must work both collaboratively and independently to develop their instructional units. Although it varies greatly depending on experience, team composition, and the time available for collaborative planning, teams can often develop new units in just a few weeks. Typically, they then revise their drafts based on how the instruction works with their students. It is also critical that time is allocated to provide the teachers with any required professional development they may need, such as instructional design, assessment planning, and training about local environmental content. Although EBE programs typically focus on investigations near the school, funding is sometimes needed to purchase materials, equipment, and other resources, or to pay for local transportation. The teachers at Seven Generations Charter School, for example, are allotted time for team collaboration three times each month on early-release days—they all report how vital this is to the successful planning of their community-based investigations.

Teachers often need instructional support and coordination while they are developing a new EBE program. Support for instructional design can come in the form of either an outside entity like SEER or other, experienced faculty members at the school. Without the help of an individual assigned to be the program leader or coordinator, teachers can easily become overwhelmed by the demands of coordinating work among teachers, community partners, and even other school staff members, such as technology staff, facility managers, custodial staff, and bus drivers. During the first year of their EIC Model implementation, the teachers at Arabia Mountain High School received ongoing support from SEER's staff. During subsequent years, this responsibility has been transferred to a team composed of an assistant principal and grade-level mentor teachers.

Even though teachers developing EBE programs are typically interested and excited about implementing community-based investigations with their students, very few of them have any technical background regarding environmental topics. Similarly, especially in medium- to large-size school districts, many teachers may live a distance away and are not very familiar with the community where their school is located. Local environmental organizations, nature centers, museums, universities, garden clubs, businesses, community groups, parents, and other individuals can be a rich source of knowledge about the local environment and community. Participating schools in Desert Sands' program had a variety of such partners that supported their efforts, and Joshua Tree National Park and the Santa Rosa and San Jacinto Mountains National Monument also provided the district with technical expertise, guides for field trips, equipment, and support for the students' community-based investigations.

Finding appropriate community partners, informing them about the school's work, and getting them involved in the program can take a substantial amount of time. Recognizing how important this work is to the success of the program, the administrators at Seven Generations Charter School have assigned a full-time staff member to work with teachers to get new community partners involved with the school as one of her responsibilities.

Teachers have the creative ability to design exciting instructional programs if they are exposed to examples of EBE programs that have been effectively implemented in other schools. Often, "painting the picture" of other successful programs is enough to encourage the teachers to kick off their own efforts. This concept was one of the main ideas behind the development of the EIC Model Demonstration School Networks within SEER's member states. Faatimah Muhammad's research into Georgia's network when she was seeking examples on which to base the design of Arabia Mountain High School's EBE program is an example of how this type of modeling can benefit other teachers and schools.

Often, the most important concern teachers and administrators have about initiating an EBE program in their school is about the difficulties they may encounter when implementing the new program. For this reason, it is very important to be prepared to respond to their concerns about potential challenges like those listed earlier.

Challenges for Districtwide EBE Program Implementation

School districts face both somewhat different challenges than individual schools in implementing EBE programs and also some of the same, school-level concerns, though on a larger scale. A decision to implement an EBE program in several schools or across a district may or may not require action by a school board. Nonetheless, it is always important to keep board members and senior district staff members informed, involved, and aware. Among the additional challenges involved with implementing an EBE program across a school district are monitoring both the efficacy of the program and student assessment, and administrative, budgetary, and program management.

School board members and district staff members usually ask different questions than school site administrators. These higher-level decision makers are usually

Challenges to Implementing EBE Programs at a District Level

Some of the challenges of and concerns about implementing EBE programs at a district level include:

- understanding the diverse perspectives of parents, community members, and local businesses regarding EBE programming;
- obtaining assurances regarding the district's authority—from a county office of education or state department of education—to implement a new program;
- adequacy of funding and other resources from existing or potential partners to support the program;
- availability of mechanisms for monitoring the progress and challenges of a program and for resolving issues that may arise, either from the perspective of administration, program implementation, or effects on students;
- possible varying levels of commitment among school site administrators and teachers for implementing and supporting an innovative program, as well as the possibility of staff changes at the schools;
- potential of the program to cause bad publicity or "political" problems in the community, region, or state;
- uncertainty about how any new program will affect the district's existing programs or be affected by changes in policy that may be made by their state department of education.

more focused on challenges related to budgets, transportation, safety, the scope and scale of any proposed program, and the level of support that exists among parents, community members, and local businesses.

These school board and district administrative concerns make it essential that an implementation plan be thoroughly grounded in concrete information backed by supporting data. Achieving this at Desert Sands Unified School District, for example, involved a series of meetings with Assistant Superintendent Darlene Dolan and Program Coordinator Ann Morales. SEER provided these administrators with the research encompassed in *Closing the Achievement Gap*,[3] a thorough description of the professional development program including the syllabus, a timetable for implementation, the program evaluation plan and associated instruments, and the credentials of SEER's team.

Since both district administrators and board members will be concerned about potential difficulties, organizations and individuals planning to promote an EBE program at this level should be prepared to respond to a wide range of implementation challenges, such as those previously listed. This type of preparation is essential to effectively engaging the school district administrators in implementing an innovative instructional program like EBE.

Challenges for Statewide EBE Program Implementation

Attempts to initiate a state-level environment-based education program confront potential obstructions that are both much larger and more complex than those at school or district levels. Although some processes are related, working with state government on educating students about the environment requires a great deal more time, funding, and understanding of the state education system. Influencing state policies and practices requires: good lobbying skills and substantial political maneuvering; thorough understanding of the roles, jurisdictions, and responsibilities of each education agency and decision maker; and an in-depth understanding of the laws and policies that constrain key educational decisions.

Although states vary, key stakeholders almost always include: governors and senior staff members in the executive branch, legislators, state boards of education, and state departments of education. Influencing the individuals in each of these bodies requires a good understanding of the factors that enter into their decision making, such as: desire to improve student achievement; interest in influencing classroom content; concerns over budgetary matters; apprehension about potential lobbying by organizations and businesses; and many others.

As the profiles of EBE programs in this chapter illustrate, similar processes are involved in planning and implementing these programs whether at the school, district, or state level: establishing goals, building networks of partners, getting permission to implement, and choosing an appropriate environmental context or set of unifying principles. The chapters that follow will outline these processes in detail.

APPENDIX (CHAPTER 4): ORGANIZATIONS THAT PARTICIPATED IN THE EEI TECHNICAL WORKING GROUPS

Environmental Organizations

- Aquatic Outreach Institute
- California Communities Against Toxics
- California Institute for Biodiversity
- Center for Resource Solutions
- Clean Water Action
- Coastkeepers
- Council for Watershed Health
- Defenders of Wildlife
- Environmental Defense Fund
- Heal the Bay
- Keep California Beautiful
- Ma'at Youth Academy
- National Wildlife Federation
- San Diego Natural History Museum
- The Nature Conservancy
- TreePeople
- Water Education Foundation

Government Agencies

- California Air Resources Board
- California Bay-Delta Authority
- California Department of Boating and Waterways
- California Department of Conservation, Division of Oil, Gas & Geothermal Resources
- California Department of Conservation, Office of Mine Reclamation
- California Department of Fish and Wildlife
- California Department of Parks and Recreation
- California Department of Pesticide Regulation
- California Department of Public Health, Division of Drinking Water and Environmental Management
- California Department of Resources Recycling and Recovery
- California Department of Toxic Substances Control
- California Department of Water Resources, Division of Planning and Local Assistance

- California Department of Water Resources, Environmental Services
- California Environmental Protection Agency
- California Integrated Waste Management Board
- California Natural Resources Agency
- Los Angeles County Department of Public Works
- Office of the Secretary, California Environmental Protection Agency
- Sanitation Districts of Los Angeles County
- State Water Resources Control Board
- U.S. Environmental Protection Agency

Industry and Private Institutions

- American Chemistry Council
- American Plastics Council
- BP America, Inc.
- California Conference of Directors of Environmental Health
- California Forest Products Commission & The Forest Foundation
- California Ocean Science Trust
- Consulting Engineers & Land Surveyors of California
- Gladstein & Associates
- Glass Packaging Institute
- Heschong Mahone Group
- HGST
- Intel
- K–12 Alliance
- Lockheed Martin Aeronautics Company
- McGuire Environmental Consultants, Inc.
- National Geographic Society
- Pacific Energy Center, Pacific Gas and Electric Company
- Sempra Energy
- The RAND Corporation

Universities

- California State University, Chico (College of Natural Sciences)
- Humboldt State University (California Geographic Alliance)
- Occidental College (Pollution Prevention Education & Research Center)
- Scripps Institution of Oceanography

- University of California, Berkeley (College of Natural Resources, Department of Environmental Science, Policy, & Management)
- University of California, Davis (Cooperative Extension in Wildlife, Fish, and Conservation Biology)
- University of California, Los Angeles (Department of Organismic Biology, Ecology and Evolution)
- University of California Cooperative Extension
- University of California, Riverside (Department of Entomology)
- University of California, Santa Barbara (Bren School of Environmental Science & Management)
- University of California, Santa Barbara (Marine Science Institute)
- University of California, Santa Barbara (Environmental Studies Program)
- University of Southern California (Keck School of Medicine)

PART TWO

Creating and Implementing an EBE Program

CHAPTER FIVE

Planning for Success

Enthusiasm for starting an exciting new environment-based education (EBE) program—or even simply a new EBE instructional unit—can be as powerful as the sensation of Herculean strength that weight lifters get when they are trying to raise a 350-pound barbell over their heads. But, like the adrenaline that gives that final boost of energy to an athlete, enthusiasm alone is not sufficient to establish a successful EBE program. Investing time and effort into creating a comprehensive EBE implementation plan is repaid many times over, in much the same way as is the work devoted to developing good weight-lifting technique.

The next five chapters are devoted to the nuts and bolts of planning, creating, implementing, and evaluating an EBE program. This chapter focuses on development of an implementation plan and other key activities that must be undertaken to get an EBE program started and sustainable. It examines the establishment of an implementation and planning team with a clear and specific statement of the program's vision, mission, and goals. The remaining chapters will introduce the major components of the program and discuss examples of how they have been executed. Whether you are a teacher interested in adding a new EBE unit to your classroom repertoire, a principal or superintendent looking to start a schoolwide or districtwide EBE program, or an environmental advocate seeking to build a statewide program or network of EBE-focused schools, the roadmaps that lead to success are very similar.

Developing a program for one classroom, a single district, or schools across a state requires thoughtful planning that focuses on the most important work and establishes clear timelines and agreement on roles and responsibilities. Planning the tasks and work assignments for creating a small EBE program or instructional unit is usually straightforward. Small programs, like a grade-specific instructional unit, can be planned by two or three people in a matter of days or weeks, depending on the unit's scope. Initiating a schoolwide program usually requires several months, because it will involve many planning sessions with representatives from all grade levels, specialist teachers, and administrators—and ideally, parents, community members, or cooperating organizations. Large-scale programs, like California's Education and the Environment Initiative (EEI), can easily involve years of work in the coordination of hundreds of individuals who represent all stakeholder interest groups. Nonetheless, whether in a single school or throughout a state, success in planning an EBE program comes only by clarifying what needs to be done, setting quality expectations for each task or work product, identifying who is responsible for the work, and establishing a timeline and budget for all of the major activities. Each of these elements of the plan helps to keep the planning team participants accountable to themselves and to all others who are depending on their work.

Although having a plan is important, involving stakeholder groups in the *process* of developing a program plan is equally important. Working together to create a program plan increases the buy-in and commitment from all organizations and individuals whose efforts are needed to implement an EBE program. People are more likely to become deeply engaged in creating or funding a program when they have participated in the program design and development process. Allowing them to share their perspectives, priorities, and skills during planning meetings also helps to build the intraorganizational and interpersonal working relationships that are so crucial to implementing a successful program. At each step, we will discuss how EBE program planners—teachers, principals, district and state staff members, community members, and partners—can include others in the work so they can build a sustainable program.

The key to successfully designing and implementing a new EBE program involves first asking and answering some basic yet very critical questions:

- Who are our stakeholders and likely partners who can help us succeed? (Whom do we need to succeed?)
- What is our mission and vision for the program? (What is our purpose, and what values will drive our program?)

- What are our goals? (What do we want to accomplish?)
- What should be the scope and scale of our program? (What should be the size of our program?)
- How do we get permission? (What backing, policies, or laws do we need to understand or develop so we can implement our program?)

While it is possible to initiate a program without first investing the time and effort in developing a thorough plan, failing to plan usually results in wasted time, funds, and perhaps most important, a lack of the energy, excitement, and momentum that are so vital to successful programs. On the other hand, good program planning often helps to avoid many of the challenges and pitfalls that might otherwise be encountered in implementing an EBE program or any other educational innovation.

BUILDING A NETWORK OF STAKEHOLDERS AND PARTNERS

Identifying and engaging a cadre of diverse stakeholders and partners, and getting them actively involved in an EBE program, is often a key to a program's long-term success and sustainability. It is equally important to realize that key stakeholders, like parents and school district administrators, can become impediments to success if they are not informed and kept up to date on the program's plans and goals. These individuals and organizations will be important assets at every stage of the program if they can be engaged in program planning—offering their expertise and technical knowledge, supplying information about potential financial and material support, helping develop and strengthen community and decision-maker involvement, or offering learning opportunities once the program is in place. (Appendix C, at the back of the book, provides examples of the wide range of roles, activities, and support that school and district teams of educators may need from various stakeholder groups and partners while implementing a typical EBE program.)

In most cases, individuals, agencies, and organizations will "buy in" to a program and provide significant support or become partners only if they have a good understanding of your program's goals, plans, and needs, and of why it is likely to succeed and who else supports it. In SEER's experience, it is only after potential partners have this depth of understanding that they will make a meaningful commitment to help implement or to become an advocate for your program. Programs may sometimes receive some level of support without detailed information, but such support is usually minor and limited to small contributions of funds or materials.

In our experience, numerous individuals, businesses, agencies, and organizations are willing to support schools implementing environment-based education programs. School- and district-level programs usually find several key interest groups—such as parents, teachers, environmental experts at agencies and universities, owners of local businesses, local elected officials, and members of service organizations. State-level programs are typically more "political," and need to engage staff members from educational and environmental agencies, state-level teacher and parent organizations, nongovernmental environmental organizations, and business and industry associations. Table 5.1 identifies the broad array of agencies, groups, representatives, and individuals who should be considered during the process of reaching out to build the program planning team and the network of stakeholders and partners.

Assembling an effective group of team members, stakeholders, and partners can require a substantial investment of time, because, depending on the scope of the program, it may involve many people. Most of this time is invested in identifying the most important stakeholder groups, determining the best way to contact and engage them, and working toward having a diverse representation of the agencies, organizations, and individuals who are most likely to contribute to, benefit from, or be affected by the program.

Depending on the time and resources available, there are many different approaches one can take to identify and communicate with stakeholders. The Larkspur Elementary School in Douglas County, Colorado, for example, announced the development of their program in the district newsletter, introducing parents and community members to the EBE instructional program they were planning to implement and inviting people to participate in informational sessions being conducted by the principal. With the scale of the program in California, several approaches to outreach were used, including: establishing an "EEI Partnership" composed of representatives from state agencies, educational and environmental organizations, and business groups; conducting statewide focus-group sessions

TABLE 5.1

Potential team members, stakeholders, and partners

Educational setting	Typical types of team members, stakeholders, and partners
Classroom, school, or school district level	• Teacher(s) • School and local school district administrator(s) And, as needed: • Parents • School facilities managers • Representatives of local, regional, and national environmental agencies and nongovernmental environmental groups • Technical experts from local nature centers, museums, and other educational facilities, such as colleges and universities • Community members, groups, and leaders • Businesses and business leaders • Local elected officials • Representatives of local and regional government agencies • Members of service organizations (Rotary Club, etc.)
State level	Representatives from: • State education agencies • State environmental and natural resources management agencies • State officials from key policy-making bodies, such as governors' offices, legislative bodies, and state school boards • State-level teacher and parent-teacher organizations • State-level associations of school board members and school district administrators • Nongovernmental environmental groups • Technical experts from colleges and universities, as well as other educational facilities, such as nature centers, museums, etc. • Business and industry organizations • Philanthropic organizations

to inform interested individuals and organizations; requesting public input on all draft materials—from the Environmental Principles and Concepts to drafts of the EEI curriculum; and conducting regional training sessions for potential implementation partners like museums, nature centers, and universities.

The specific group of team members, stakeholders, and partners needed to implement an EBE program depends on the specifics of each program: scope and scale; stage of development; needs for scientific and technical expertise; and access to funding and implementation resources. The interests and potential roles of each group member should be taken into account during the process of developing a support network, whether for a large or a small EBE program, and reviewed as the program develops and grows, so that the network's composition can be adjusted to meet changing needs.

The involvement of team members, stakeholders, and partners can provide a wide array of important support and resources to an EBE program, whether it is just beginning or has already been successful for several years. The participation of these individuals, governmental agencies, environmental organizations, businesses, community members, and technical experts is vital to establishing a program that: is instructionally effective and scientifically and historically accurate; presents balanced information about environmental issues and is unbiased politically; and has the resources available to undertake all necessary program activities.

The roles of team members, stakeholders, and partners will vary during the planning and implementation phases of an EBE program. Needs change, levels of interest grow and shrink, and different types of expertise are needed as the program moves from planning to early stages of implementation and to evaluation and an ongoing process of continuous improvement.

ENVISIONING SUCCESS

Whether it is called "dreaming the dream" or simply "creating vision and mission statements," it is very important to the planning process of a successful EBE program. This is especially true since these programs require the backing, support, and involvement of so many different stakeholders. Clear statements help to keep everyone focused on the same endpoint.

Discussing and reaching agreement on the vision and mission statements for a new EBE program is a process that is sometimes taken for granted. Many individuals and organizations assume that, because they use common terminology, they have a shared understanding of what they want to accomplish. Since this is not usually the case when multiple organizations are involved, in-depth, substantive discussions are an essential step toward ensuring that all participants have the same understanding of what they are trying to achieve by creating a new program.

Support from Team Members, Stakeholders, and Partners

The team members, stakeholders, and partners associated with EBE programs in various parts of the country have provided four types of critical support: technical, financial, political, and instructional. Some examples of this support include:

- sharing technical knowledge and professional expertise with teachers and students about content that the teachers may not be familiar with, such as scientific knowledge, environmental content, or community history. For example, Brundrett Middle School in Port Aransas, Texas, a school in Texas's Strands Program, obtained technical support and scientific guidance from staff members at the University of Texas Marine Science Institute and Texas A & M University Center for Coastal Studies, who taught the students about local environmental concerns and coastal and marine habitats and how to correctly collect and analyze scientific data. The Port Aransas Preservation and Historical Society provided class presentations, local field trips, and student materials for the study of historical patterns of resource use in their community.

- supplying implementation resources and financial support, such as fund-raising and/or providing funds for instructional materials, supplies, or items like transportation. For example, Armuchee Elementary School in Rome, Georgia, received: a total of about $10,000 and four years of professional development and technical support with a value of approximately $20,000 from the Georgia Environmental Protection Division for developing their EIC Model program; mature plants and seeds from the State Botanical Garden of Georgia to establish a bog garden; and empty barrels from the local Coca-Cola bottler to kick off a student-led water conservation project.

- developing community and political support, such as championing the program with educational administrators, local and state school board members, and parent-teacher organizations in order to garner support or encourage the development of policies that allow the implementation of an EBE program. For example, establishing Georgia's statewide network of EIC Model schools required approval from Georgia's State Board of Education, and Lynn McIntyre, representing the North Fulton Council PTA, organized an "informational campaign" for the State Board of Education on behalf of the program and brought in experts like Ann Bergstrom from the Chattahoochee Nature Center to testify on its behalf.

- offering learning opportunities and instructional support, such as participating in student instruction in the classroom, leading instructional field trips, and supporting instructional activities outside of the classroom. For example, students at Concrete Middle School in Concrete, Washington, participated in a field study at the North Cascades National Park so they could learn from park staff and apply what they learned to building a nature trail by the school. They also met with experts from the Washington Department of Fish and Wildlife in preparation for developing their own research projects.

Vision and Mission Statements

Developing the vision statement for a program is usually the first step in this process. These statements are intended to present an idealized view of what an EBE program can and should be. These are typically written at a "high level" that presents a long-term view for the future—more related to the values the program will promote than the pragmatic aspects of the endeavor.

As large as California's Education and the Environment Initiative is, its vision statement is quite brief: "Increasing Environmental Literacy for K–12 Students . . . Because the Future Is in Their Hands." This simple statement encompasses the key features of an effective vision statement, explaining in just a few words the purpose of the program and the underlying values of the agencies and individuals responsible for its implementation.

Mission statements are somewhat different, in that they should present a pragmatic view of the EBE program and describe the core purpose of the program, its primary objectives, and how success will be measured. The mission statement for Desert Sands Unified School District's EBE program provides a useful example. It reads:

> Our mission is to provide all students with a Science/Math/Technology curriculum emphasizing Environmental Studies. With the assistance of committed staff, families, and community members, all students will gain knowledge, awareness, and critical thinking skills through the scientific study of the natural world and its multifaceted and interrelated systems.

Desert Sands' statement makes it clear that the program's core purpose is to help students understand the natural world and its "multifaceted and interrelated systems." The program's primary objective is to achieve this purpose through a curriculum that emphasizes environmental studies. The mission statement also speaks to the fact that progress is intended to be measured by the students' gains in knowledge, awareness, and critical thinking skills.

Typically, vision and mission statements are initially developed by a program's founders. They are dreaming about what they would like their EBE program to look like. The founders of Seven Generations Charter School, for example, met many times over the two years before the school was established to work on defining their mission and a vision of what they would want the school to be. These statements have been refined over the past several years—a process that is typical as new stakeholders become involved in a program. The vision and mission statements for Seven Generations Charter School's EBE program read:

Vision Statement

- An integrated learning experience that emphasizes sustainable living practices
- An experiential, constructivist approach to education that encourages hands-on learning in the community as well as in the classroom
- A culturally rich atmosphere that celebrates the expansive world around us
- An environment of academic excellence that taps into the creativity and uniqueness of each child and that fosters mutual respect

Mission Statement

The Seven Generations Charter School is an academically rich educational community creating generations of stewards who embrace our world and each other. All members of the Seven Generations Charter School community are committed to a public education alternative that promotes sustainability and citizenship with an interdisciplinary, individualized, project-based curriculum.

Teachers usually adapt the overall program vision and mission statements to their individual EBE instructional units. For example, "Fantastic Forests," one of the EIC Model units developed by Lisa Fritz, Amanda Cossman, and Louise Moyer, the second-grade teaching team at Seven Generations Charter School, includes features from both the vision and mission statements in their plan. (See appendix: "Instructional Framework of Seven Generations Charter School's Second-Grade 'Fantastic Forests' Unit" at the end of this chapter for a summary of the instructional framework of this unit.) At the unit level, the vision statement provides a sense of the values of the school and teachers as they implement each unit. The vision of "Fantastic Forests" is a good example, stating, "In all we do, we foster respect and responsibility for ourselves, for one another and for the natural systems on which all life depends."

Unit-level mission statements, like those for the overall EBE program, should be pragmatic and clearly state their primary objectives. "Fantastic Forests," for example, identifies the mission as:

> Working together with the community, students, and teachers at Seven Generations to investigate and discover their connection to our local natural and social environments. We engage in student-driven inquiry which seamlessly blends standards-based academic knowledge across disciplines with hands-on learning, meaningful community-based service, and reflection.

The fifth-grade "Water Works" unit developed by teachers at Jackson Elementary School in Altadena, California, provides another example of a unit-level vision and mission statement. (See appendix: "Instructional Framework of Jackson Elementary School's Fifth-Grade 'Water Works' Unit" at the end of this chapter for a summary of the instructional framework of this unit.) It states:

> Our staff and community will collaborate to enrich the academic environment for our students by incorporating standards-based integrated interdisciplinary instructional strategies and learner-centered constructivist approaches. The students and instructional team will develop community-based investigations that will result in service to our community. As our students strengthen both their independent and cooperative learning skills, they will achieve higher academic standards, recognize connections between learning and the real world, become involved and caring citizens, positively influence the environment, and develop a sense of pride in themselves and their community. Ultimately, students will recognize the positive effects that they can make on the environment by becoming more aware of water as a resource, their water usage, and the need for water conservation.

Developing vision and mission statements can require a series of meetings, considering that they should represent the perspectives, needs, and underlying goals of all the stakeholders. Not making this investment of time and effort can result in setbacks to program development. For example, partners and key funding can easily be lost if the participants do not achieve consensus on these guiding statements at an early stage of program planning.

Setting Goals

Successful EBE programs do not just happen once the vision and mission statements are in hand. They require the setting of specific goals, a process that by its very nature involves digging deeply into the question "What are we trying to accomplish?" Undertaking this stage of planning requires a substantial commitment of time, because the most beneficial goals are specific, relevant, attainable, and measurable. These attributes will then drive everything, including identification of tasks, budgets, timelines, and, ultimately, measures of success.

In most EBE programs, the goals are focused on an educational concern that the participants are seeking to resolve—this is at the core of all the goals for their program. Some of the goals that are common to many EBE programs include improving academic achievement, increasing student engagement, decreasing classroom discipline problems, developing career and life skills, and developing students' "environmental literacy" and behavior. The four programs described next present examples of several of these different goals.

Lifting academic achievement was the primary goal of Principal Mary Janda when she began an EBE program at the Concrete Middle School in Concrete, Washington. As the school had some of the lowest achievement scores in their county, Janda knew she had to do something. Looking to secure federal school improvement funding, she asked parent volunteer Kathleen Howe to help find an educational model that could serve the students in their small community. Living in the dense forest of the mountains of the North Cascades National Park, Janda and Howe felt their students would find an environment-based education model engaging. They brought their ideas to District Superintendent Marie Phillips and the school board, who agreed with their goal and the plan. With the clear goal of improving student achievement defined, the district applied for federal funding and was awarded a grant of $180,000 for the three-year program.

In planning for their Voluntary Public School Choice grant, Desert Sands Unified School District, in California's Coachella Valley, started with a clear goal: to better engage students attending seven of the district's low performing schools. They did not start out with a focus on the environment; it was not an intrinsic part of their overarching goal. However, the message they received from parents and community members was that they wanted their students to participate in an education program centered on science and the environment. The district's program planners, Superintendent Wilson and Assistant Superintendent Dolan, merged these ideas and arrived at the goal they brought to the school board. "We explained," said Dolan, "that the goal of the program was to develop a comprehensive K–12 school choice program that will increase student engagement by offering them opportunities to participate in a science- and environment-focused program."

Similarly, in South Carolina, State Superintendent Inez Tenenbaum was looking for a creative instructional strategy—any instructional strategy—that would increase student engagement and thus help her school districts decrease the dropout rate of boys transitioning from middle school to high school. Results were seen within the first year of implementing EBE in ten South Carolina schools—Dr. Tenenbaum said school administrators reported improvements in student attendance, a decrease in discipline referrals, increased parental and community interest, and greater support of student learning. Unfortunately, in spite of the success of this program, after Dr. Tenenbaum's retirement, the incoming state superintendent changed many of the goals of the department of education. Changes of leadership and associated goals are elements that will require adjustments at all levels of implementation, from school through state.

In Rigby, Idaho, for example, eighth-grade social studies and science teachers Laron Johnson and Alvin McKenna at Rigby Junior High School felt it was important that students learn about systems thinking and environmental sustainability. Between them, they sketched out their ideas for creating an environmental field studies program. They took their ideas to Principal Sherry Simmons, discussed what their program might look like, and agreed on an overarching program goal of creating a standards-based environmental learning program to give students authentic learning experiences and opportunities to participate in a variety of environmental service-learning projects. This was the beginning of Rigby's Environmental Field Studies program.[1] A few years later, the school was invited to join a network of eleven EIC Model schools from across Idaho, Montana, and Wyoming within the Greater Yellowstone Ecosystem. This gave Johnson and McKenna the opportunity to expand the program they had started and to bring in math and English teachers to strengthen their interdisciplinary team. Demonstrating that the goal of increasing environmental literacy was as much hers as theirs, Principal Simmons joined them at an EIC Model professional development institute.

At whatever level—school, district, or state—EBE teams should work to define meaningful, achievable goals that will guide their work. Throughout the implementation process, they can then measure their progress toward these goals and make adjustments to their program to ensure that they are always operating in ways that will allow them to succeed.

DETERMINE THE SIZE OF YOUR PROGRAM

Environment-based education programs can be successfully implemented on a wide range of scales, from individual classrooms to statewide. Achieving the program's goals, however, depends on matching the scale and scope of your plans to the resources that are available for implementation—the adage "Don't bite off more than you can chew" applies to this situation.

Deciding how big a program should be requires understanding the level of demand and identifying the resources that will be needed for implementation. The key questions include: how many students, teachers, classrooms, or schools will be involved if the program is initiated? and what types of resources will be needed, how much of each, and where might these be obtained?

If funding and support are limited, environment-based instructional units can be implemented by an individual teacher in one classroom—usually with no or very low costs, and often augmented with contributed services. Science teacher Loris Chen, for example, designed a small EBE program at New Jersey's Eisenhower Middle School in Wyckoff to give her seventh- and eighth-grade students an opportunity to apply what they were learning about the sun, plants, and the heat island effect. Considering the constraints on available resources, Ms. Chen decided that she and the students should identify a classroom-sized environmental topic—the overheating of the school's classrooms by direct sunlight through the windows—as their context for learning selected earth and life science standards. The students used what they had learned in science as the basis for developing a plan to resolve the problem. Ms. Chen's students were extremely fortunate, because they received the help of twenty-two local businesses, and financial support from two foundations and a community group. With these resources, they were able to undertake the environmental service-learning project they had designed—landscaping the school's courtyards with plants and trees that would provide shade to cool the overheated classrooms. A single-classroom approach is not ideal for EBE, because these programs are most effective when implemented across several grades

Program Implementation Costs

The costs of implementing EBE programs vary greatly, based on both the size of the program and its implementation plan, so there are not any good rules of thumb. The examples below demonstrate the broad range of potential costs at school, district, and state levels.

School

- Many individual schools have initiated programs by allocating teacher planning time during the day, but without any cash outlay.
- Red Oak Middle School, in Red Oak, Iowa, was allocated about $4,000 over the first two years by the Iowa Department of Education and received approximately $4,000 of in-kind services for professional development and program evaluation from other sources.
- Seven Generations Charter School in Emmaus, Pennsylvania, invested approximately $130,000 over the first four years for external professional development and technical support, as well as for other expenses such as substitute fees, funding for which was part of the school's general operating budget.

District

- Desert Sands Unified School District, in California's Coachella Valley, expended approximately $9 million over a five-year period for the six schools and the district staff members participating in the program, funding for which came from federal Title I funds and the State of California.

State

- Georgia's EIC Model Demonstration School Network ultimately involved fourteen schools and a total cost of cost approximately $250,000, or about $4,000 per school each year over five years, funding for which was provided by the State of Georgia, with cash and approximately $140,000 of in-kind services for technical support, professional development, and program evaluation from other sources.
- California's Education and the Environment Initiative, through all stages of development and creation of the EEI curriculum, spent over $10 million, not including printing, distribution, or professional development. These funds were all provided as part of the State of California's annual budget.

and disciplines. Nonetheless, many teachers have successfully developed programs like Ms. Chen's with little or no outside financial support.

Not surprisingly, developing an EBE program to benefit an entire school takes more time, effort, and cooperation than single-classroom programs. Schoolwide efforts usually require a commitment from the principal and a majority of teachers, and take organizing these individuals into a planning group that can work together for six to eighteen months to complete the initial phases of program planning. Developing new instructional units then continues as part of the ongoing work of grade- or discipline-level teams.

Principal Anita Stewart and her successor at Armuchee Elementary School in Rome, Georgia, made this major investment of resources as part of a statewide initiative to develop model schools. Seeing it as a way to tie together best teaching practices, Stewart selected a core group of teachers to work with their community partner, Arrowhead Environmental Education Center, in designing the program. The interest of the administrators at the school and at the Floyd County Office of Education was so deep that, over the program's first four years, they had all school faculty members participate in SEER's four-day EIC Model professional development program, and also gave them planning time to create many new instructional units. After all this hard work by individual teachers, requiring numerous team meetings and partnerships with community members, the Armuchee team had built an impressive schoolwide third- through fifth-grade EBE program. The teachers and students chose the Armuchee Creek watershed as the context for the program, and it served them well across all three grades through studies of the forest behind the school, the nearby three rivers area, and many local environmental issues.

Additional support came from Arrowhead Environmental Education Center and the county office of education in the form of buses and staff support for numerous field studies. The community also joined in, with diverse volunteers contributing rich information about the local environment.

Some of the largest EBE programs are statewide in scale. California has its EEI program, and Texas and other states have developed demonstration networks of schools that have implemented EBE programs. In 2002, the Texas Education Agency decided to take a statewide approach, and created Texas Strands, a program intended to create a network of schools to model the EBE strategy for schools implementing their new science standards. The agency had previously developed other networks of schools to demonstrate effective instructional approaches to administrators and teachers across the state by giving them an opportunity to visit, observe, and learn from these schools. Located from Dallas to the coast and east to west, the seven schools chosen reflect the demographic and geographic diversity of Texas. Educators from these schools participated in a professional development institute and received ongoing technical support from SEER, the Southwest Education Development Laboratory, and the Lower Colorado River Association. Each of the schools was represented by interdisciplinary teams who were creating EIC Model units that they could use to integrate teaching of state standards across multiple disciplines.

In addition to financial support from the Texas Education Agency, these schools' programs were supported by one or more local partners ranging from universities to community organizations and local environmental groups. This support was especially important in Texas, where the great distances between the agency coordinators in Austin, the technical support team at SEER, and the participating schools made it extremely time consuming and costly for the state representatives to visit them on a regular basis.

Determining program needs and matching them with realistic, sustainable levels of funding and resources is important to achieving EBE program success. Like a small business without sufficient start-up funds, programs can fail if they lack the necessary support and resources. At the same time, new programs can be damaged by having too much funding if it forces them to scale up their efforts before they have sufficient infrastructure.

GETTING PERMISSION

Time and effort should be devoted to the process of seeking and securing "permission to implement" a new EBE program from appropriate decision makers. Simply put, over both the short and long term, having such support can "make or break" any innovative education program.

The policies that control education at the local, state, and national levels are ever-moving targets. Speculating about the changes in direction from one year to the next is difficult, because there are so many different perspectives on what constitutes "good education" and how best to make it happen. This situation is complicated further by the fact that critical policy decisions are made at every level from principals at individual schools to the president in the White House.

The problems associated with a constantly changing educational-policy climate are further exacerbated by a multitude of other, seemingly endless changes in local and state education systems. These changes encompass both major and minor items, including: changes in school, district, and state administrations; revision of state standards; implementation of new state- and district-mandated student assessments; increasing numbers of students in classrooms; introduction of new instructional programs and materials; increasing demands on teachers and school administrators; changing funding levels; and a variety of others.

The rapidity with which all of these changes occur can be daunting to those working to implement an educational innovation like an EBE program, whether at a school, district, or state level. It makes it extremely important to the designing and implementing of a program that permission to implement be obtained—whether this permission is verbal (although not the ideal) or in the form of a formal decision, policy, or state law.

Although gaining formal permission in any of these forms is not a guarantee, it is one of the most important means of increasing the likelihood of the long-term success and sustainability of an EBE program. Other benefits of investing the effort to gain appropriate levels of permission include:

- Increasing the chances that a program will continue after changes of administration or staff;
- Enhancing the program's credibility, thereby expanding opportunities for funding, since many grant programs require indications of formal policy or legal standing;
- Enlarging the pool of potential agency, community, or business supporters;
- Deepening decision makers' understanding and the public's awareness of a program.

Permission to establish an EBE program can take several forms and can be given by decision makers at a variety of levels: from an oral understanding with a principal; to a policy established by a school, district, or state department of education; all the way to the highest levels—getting a new state or federal law enacted. The simplest types of permission can typically be gained the most easily, but these are generally much weaker and less durable than the higher levels of permission.

Many programs get started on the basis of a simple oral authorization, perhaps taking the form of a principal saying, "Go ahead and try it, let's see how it works." That is just what happened when June Burton, one of the third-grade teachers at Twin Lakes Elementary School in El Monte, California, approached Principal Angelica Sifuentes-Donoso to ask if she could expand her recycling program to other classrooms and grades. Burton engaged several of her colleagues, and working together over the next two years, they expanded their efforts into a broader, standards-based EBE approach that extended the focus well beyond recycling and into English language arts, math, science, and history/social science in kindergarten and the first, second, third, and fourth grades.

Obtaining a more formal authorization to implement or expand a program—such as permission from a local school board or superintendent—can be complicated and time-consuming, but it is usually much more enduring. Noel Buehler, superintendent of the Oak Grove Union School District, in Santa Rosa, California, already had several teachers involved in an EBE program at the district's elementary school when he decided to seek the support of his school board to commit to expanding their program. He drafted a board resolution that authorized him to apply for a grant from the State of California, and he made a presentation to the board that described his plans for developing a program that would "involve [student] participation in integrated waste management, energy and water audits, and other resource conservation efforts that focus on identified school and community needs." The school board approved his proposals for developing a standards-based EBE program, and students continue to reap the benefits of this program ten years later, even with a new superintendent in place.

The Pennsylvania State Board of Education's 2002 approval of academic standards for "Environment and Ecology" did more than just give educators permission to teach about the environment—it made it a requirement. As an outgrowth, the Pennsylvania Department of Education defined their own task by stating: "It is our responsibility to develop a citizenry that is aware of and concerned about the total environment and has the knowledge and skills to work toward solutions to current problems and the prevention of new ones . . . [This will allow] students to understand, through a sound academic content base, how their everyday lives revolve around their use of the natural world and the resources it provides." Subsequent to the establishment of this goal, Dr. Patricia Vathis, the environment and ecology curriculum adviser for the department, has been working with school districts, schools, and teachers through state-sponsored professional development and technical assistance programs to help them implement these standards. The significance and permanence of fully integrating these standards into every school is further enhanced by the fact that students' achievement of the standards is tested as part of the annual Pennsylvania System of School Assessment.

The level at which permission should be sought depends on a program's specific goals and its intended scope and scale, as well as on the knowledge and support available from the stakeholder and partner network. Although these will vary significantly from one situation to the next, some of the key decision makers and leaders at each of the levels include:

- **School.** Teachers, school council, parent-teacher organization, and principal
- **School district.** Director of curriculum and instruction, superintendent, and school board
- **State.** Senior staff members and administrators at state department of education, state superintendent, state board of education, state legislature, and governor's office

Douglas County School District Puts It All Together

Soon after arriving at Colorado's Douglas County School District in 2010, new superintendent Dr. Elizabeth Celania-Fagen and her team began developing a new strategic plan for the district's eighty-nine schools. She wanted to create a strategy that would achieve the district's goal to "reimagine and reinvent American education so our students are prepared to compete on a world stage for any college and career pathway of their choice." Through this planning process, they developed three district-level priorities: providing each child maximum opportunities for success through school choice; achieving sustainable learning for the twenty-first century through world-class education; and defining and measuring what matters most as the means of monitoring system performance.

(continues)

With this vision and these goals in place, Dr. Fagen empowered principals across the district to seek out instructional models that would best fit their students, schools, and communities. With this "permission" and the encouragement and backing of other district administrators, Michael Norris, principal of Larkspur Elementary School, began the process of reviewing a variety of instructional models that would help his school achieve the district's key priorities, including sustainable learning so that the "most important information [is taught] in a way that can be retained, that stretches across content areas, and that incorporates the tools our young people will be required to use in the workforce."

After reviewing many different instructional strategies and visiting the Seven Generations Charter School in Emmaus, Pennsylvania, Mr. Norris chose to implement the EIC Model at Larkspur. Since making that decision in early 2013, he has been working actively to discuss and engage all of the school's stakeholder groups, including district administrators, teachers, parents, and community members.

Superintendent Fagen expressed her strong support when she said, "This really is an amazing idea. Our students will have the opportunity to explore, hypothesize, test, and defend positions based on real data—data owned by them. It is a model that provides the kind of learning experiences that our students need to be successful—the most successful in the world."

The next four chapters describe and discuss the major instructional components for designing and implementing an EBE program: choosing an environmental context; selecting content standards and defining learning objectives; developing instructional units and materials; and assessing student learning and program effectiveness.

APPENDIX A (CHAPTER 5): INSTRUCTIONAL FRAMEWORK OF SEVEN GENERATIONS CHARTER SCHOOL'S SECOND-GRADE "FANTASTIC FORESTS" UNIT

Vision and Mission

"In all we do, we foster respect and responsibility for ourselves, for one another and for the natural systems on which all life depends. Working together with the community, students, and teachers at Seven Generations to investigate and discover their connection to our local natural and social environments. We engage in student-driven inquiry which seamlessly blends standards-based academic knowledge across disciplines with hands-on learning, meaningful community-based service, and reflection."[2]

Environmental Context

Natural Systems

The forest environment in and around Emmaus, Pennsylvania, including the trees, plants, soil, wildlife, and decomposers.

Human Social Systems

Emmaus community (social systems in and around Emmaus, Pennsylvania), including people, businesses, paper mill, sawmill, tree nursery, furniture makers, railroad, and local government.

Natural and Human Social Systems Interactions

We use forest resources for many reasons, and our choices to do so have a variety of effects on our forest environment. The ways in which we utilize forest resources will have both positive and negative effects on the forest's future. These interactions affect humans, wildlife, plants, and the rest of the forest ecosystem, and local and regional social systems.

Standards

English Language Arts

Pennsylvania Standards (selected examples)[3]

1.2.2.D. Make inferences from text when studying a topic (e.g., science, social studies) and draw conclusions, citing evidence from the text to support answers.

1.5.2.B. Develop content appropriate for the topic: Gather and organize information, incorporating details relevant to the topic; and Write a series of related sentences or paragraphs with one central idea.

ELA Common Core (selected examples)[4]

RL.2.1. Ask and answer such questions as who, what, where, when, why, and how to demonstrate understanding of key details in a text.

RI.2.4. Determine the meaning of words and phrases in a text relevant to a grade 2 topic or subject area.

W.2.2. Write informative/explanatory texts in which they introduce a topic, use facts and definitions to develop points, and provide a concluding statement or section.

W.2.7. Participate in shared research and writing projects.

Math

Pennsylvania Standards (selected examples)[5]

2.7.2.E. Ask and answer questions about predictions and actual outcomes based on data.

2.5.2.A. Develop a plan to analyze a problem, identify the information needed to solve the problem, carry out the plan, check whether an answer makes sense, and explain how the problem was solved in grade appropriate contexts.

Math Common Core (selected examples)[6]

2.MD.1. Measure the length of an object by selecting and using appropriate tools such as rulers, yard-sticks, meter sticks, and measuring tapes.

2.MD.4. Measure to determine how much longer one object is than another, expressing the length difference in terms of a standard length unit.

2.NBT.2. Count within 1000; skip-count by 5s, 10s, and 100s.

2.NBT.6. Add up to four two-digit numbers using strategies based on place value and properties of operations.

Science and Technology and Engineering Education (selected examples)[7]

3.1.2.A3. Identify similarities and differences in the life cycles of plants and animals.

3.1.2.A5. Explain how different parts of a plant work together to make the organism function.

Environment and Ecology (selected examples)[8]

4.1.2.A. Describe how a plant or an animal is dependent on living and nonliving things.

4.1.2.D. Identify differences in living things (color, shape, size, etc.) and describe how adaptations are important for survival.

4.3.2.B. Identify products and by-products derived from renewable resources.

Geography (selected examples)[9]

7.3.2.A. Identify the effect of local geography on the residents of the region (food, clothing, industry, trade, types of shelter, etc.).

7.4.2.A. Identify how environmental changes can impact people.

Learning Objectives (selected examples)

Lesson 1

- Students will know that trees are made up of parts with specific functions.
- Students will write a nonfiction booklet that shows their understanding of the structure of a tree and how the different parts of a tree help the tree function.
- Students will use observation skills to draw a picture of a tree depicting in detail different parts.
- Students will describe a chosen tree using personal observation and investigation and organize information about the tree.
- Students will diagram the life cycle of a tree and show their understanding of that life cycle and the tree's role in the ecosystem throughout its life.
- Students will add and graph acorns found at a white oak tree in the meadow over the course of five weeks.
- Students will add the total number of acorns found over the course of five weeks.

Lesson 2

- Students will identify three local trees by the size and shapes of their leaves.
- Students will understand how leaf shapes, sizes, and other characteristics vary from plant to plant by creating a field guide with at least three leaves.
- Students will identify a compound and simple leaf.
- Students will identify three local tree species by their leaves.

Lesson 3

- Students will create a model of the forest that depicts its three layers.
- Students will identify objects from the forest, state the layer the object comes from, and explain its importance to the forest.
- Students will draw and label the layers of the forest system and tell the importance of each layer to the survival of the forest ecosystem.

Lesson 4

- Students will simulate how trees compete for their essential needs.
- Students will describe ways in which animals and plants depend on trees for survival and in turn influence the tree.
- Students will describe the stages of decomposition of one object.
- Students will research one animal found in the forest and describe the four elements needed for its survival.
- Students will put seeds in groups of ten and skip-count to find the total.

Lesson 5

- Students will examine various products and determine which ones are made from trees.
- Students will determine how products made from trees make our lives easier.
- Students will create a how-to book that describes how one product is created, from tree to store.

Lesson 6

- Students will explore a variety of jobs that are directly related to forest resources.
- Students will describe how various professionals work together to care for forests.

APPENDIX B (CHAPTER 5): INSTRUCTIONAL FRAMEWORK OF JACKSON ELEMENTARY SCHOOL'S FIFTH-GRADE "WATER WORKS" UNIT

Vision and Mission

"Our staff and community will collaborate to enrich the academic environment for our students by incorporating standards-based integrated interdisciplinary instructional strategies and learner-centered constructivist approaches. The students and instructional team will develop community-based investigations that will result in service to our community. As our students strengthen both their independent and cooperative learning skills, they will achieve higher academic standards, recognize connections between learning and the real world, become involved and caring citizens, positively influence the environment, and develop a sense of pride in themselves and their community. Ultimately, students will recognize the positive effects that they can make on the environment by becoming more aware of water as a resource, their water usage, and the need for water conservation."[10]

Environmental Context

Natural Systems

Natural water system and sources, including the water cycle, groundwater, rivers, and runoff from Sierra Nevada mountain range. "Ecosystems," including natural and human-made habitats: "green" areas like the campus lawns, gardens, and playing fields near school; local parks, habitats, and areas around students' homes.

Human Social Systems

Education system, including: students, teachers, administrators, families, and community members. Water transport system, including: reservoirs, water supply lines, storm drains, sewage lines, and water treatment facilities. Economic system, including: school budgets, community budgets, and taxes. Transportation system, including: sidewalks, roads, cars, and buses.

Natural and Human Social Systems Interactions

- Water and soil from campus areas flooding into the school building, making a classroom unsafe and increasing costs to the school and community
- Water from the natural water system being used by the school and community
- Soil loss due to erosion, and damage to the vegetated areas on campus including lawns, gardens, and playing fields
- Excess water from the water system affecting the "green" areas on campus and entering the storm drains, increasing costs to school and community

Standards

English Language Arts/Literacy Common Core State Standards (selected examples)[11]

RI.5.7. Draw on information from multiple print or digital sources, demonstrating the ability to locate an answer to a question quickly or to solve a problem efficiently.

W.5.2. Write informative/explanatory texts to examine a topic and convey ideas and information clearly.

SL.5.5. Include multimedia components (e.g., graphics, sound) and visual displays in presentations when appropriate to enhance the development of main ideas or themes.

Mathematics Common Core State Standards (selected examples)

5.MD.A.1. Convert among different-sized standard measurement units within a given measurement system, and use these conversions in solving multi-step, real world problems.

5.MD.B.2. Make a line plot to display a data set of measurements in fractions of a unit. Use operations on fractions for this grade to solve problems involving information presented in line plots.

5.MD.C.5. Relate volume to the operations of multiplication and addition and solve real world and mathematical problems involving volume.

Science Standards (selected examples)[12]

5.3.d. Students know the amount of fresh water located in rivers, lakes, underground sources, and glaciers is limited and that its availability can be extended by recycling and decreasing the use of water.

5.3.e. Students know the origin of water used by their local communities.

History/Social Science Standards (selected examples)[13]

5.1.1. Describe how geography and climate influenced the way various nations lived and adjusted to the natural environment, including locations of villages, the distinct structures that they built, and how they obtained food, clothing, tools, and utensils.

5.8.4. Discuss the experiences of settlers on the overland trails to the West (e.g., location of the routes; purpose of the journeys; the influence of the terrain, rivers, vegetation, climate; life in the territories at the end of these trails).

Learning Objectives (selected examples)

- Analyze literature about water conservation and look for ways that they can personally save water at home and on the school campus.
- Create posters in small groups showing different ways that students can conserve water.
- In an informal discussion, describe ways that they and their families are wasting water.

- Describe how water is supplied to their school and how it is utilized at school.
- Compute how much water could be saved by a student in a day.
- Explain that the water cycle is a closed cycle that water is continuously moving through.
- Identify the areas where you can observe the water cycle as a whole.
- Describe how we can use both saltwater and freshwater.
- Draw a map of the path the water takes to make it to the students' school and homes.
- Identify the water cycle and how it interacts with their water usage.
- Describe why earlier civilizations chose to live in certain areas based on the water cycle.
- Describe the flow of water into the school and the flow of wastewater out of the school and the community at large.
- Identify interactions among the school community and the natural systems in and around the school.
- Make predictions about the effects of natural and social interactions in and around the school community.

CHAPTER SIX

Choosing an Environmental Context

Parents and teachers alike have long understood that context gives meaning. Young children learn the sounds and meanings of their first words in the context of the language spoken by their families and peers. Learners of foreign languages quickly realize that they learn the meaning of new words most readily when they are used in the context of words they already know.

Context is important not only to the learning of languages, it is equally significant as the basis for studying and learning everything from math to science to history, geography, music, and art. "Knowledge is created and made more meaningful by the context in which it is acquired."[1] A well-defined context reduces the need for students to ask the question "Why do I have to learn this?"

Choosing appropriate environmental contexts is the first, and often the most important, step in the process of planning and creating an environment-based education (EBE) program. As illustrated in figure 6.1, this choice influences the development of every other aspect of the program, including decisions about standards, learning objectives, instructional materials, and assessments—not to mention the character of the interdisciplinary projects and community investigations that are at the heart of EBE programs. These decisions directly influence a program's potential for benefiting students, whether that is measured through standards-based achievement or students' engagement in their own education.

This chapter discusses and provides examples of two distinct types of environmental contexts: one focused on the local environment and community, and the other centered around "big" environmental ideas. It distinguishes between the idea of using a particular location as simply a site at which students can study nature

FIGURE 6.1

Choices of context and instructional design

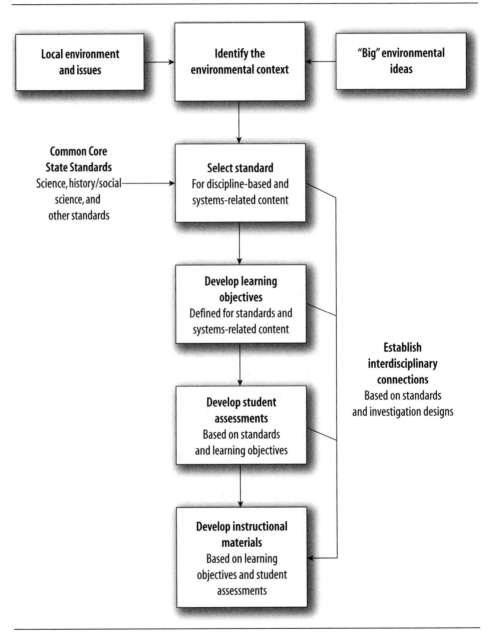

and the larger notion of environmental contexts based on the interactions of natural and human social systems. Since different environmental contexts are developmentally appropriate for different age groups, it then examines how and when teachers can involve students in selecting an environmental context as the focal point for their studies.

LOCAL ENVIRONMENT VS. BIG IDEAS

Standards-based instruction can be constructed around two different types of environmental contexts—one focused on the local environment and community, and the other centered around one or more big environmental ideas. In this sense, the phrase "big environmental idea" refers to a set of understandings that can be broadly applied, rather than to the scale of a particular environmental topic. For example, global climate change is a large-scale and very important topic, but in and of itself, it is not an overarching principle that can be applied to many situations. Two examples of big environmental ideas are California's Environmental Principles and Concepts, which can be applied globally, and Desert Sands' districtwide organizing questions, which are conceptually wide reaching even if studied only in the context of the Coachella Valley (see chapter 4).

Either of these contexts can provide the framework within which to build an educationally effective program as long as key elements such as level of student engagement, breadth of opportunities to learn about systems thinking, and the potential for improving students' standards-based achievement are taken into consideration. Each approach offers distinct advantages.

Programs that center on the local environment and concerns within their communities actively engage students by offering real-world examples about places they care about and can observe directly. Schoolyards and gardens, community parks, nature centers, undeveloped natural areas, local lakes, creeks, rivers, farmers' fields, and even vacant lots across the street from the school are among the many locations that can be developed into a fully defined local environmental context. In addition to producing a high level of student engagement, studying the local environment helps students build a "sense of place" and the perception that what they are studying in school has a meaningful connection to their daily lives. Using a local environmental context usually allows students to more readily and inexpensively visit the types of locations they are studying, and gives them greater opportunities to directly observe and learn about the natural and human social systems that exist in their community. These local contexts also allow them to analyze real-world problems and participate in meaningful environmental service-learning projects. The EBE programs at Seven Generations Charter School in Emmaus, Pennsylvania, and at Jackson Elementary School in Altadena, California, are examples of programs that use the local environment as their contexts. These examples and others will be discussed in depth in the chapters that follow.

Environment-based education programs centered on one or more big environmental ideas, like California's Environmental Principles and Concepts (see appendix A at the back of the book), on the other hand, provide students with the opportunity to examine larger, more general concepts that can allow them to create a bigger-picture understanding of the world around them. Rather than engaging students in investigating only within their local communities, this type of context helps them make broader connections and discover how these ideas apply to local, state, and global issues. For example, one of the Education and the Environment Initiative (EEI) curriculum units uses California's Environmental Principle II, which concerns the influence of human communities on terrestrial, freshwater, coastal, and marine ecosystems, as the context for studying how Britain's Industrial Revolution affected the natural and human social systems in the 1800s. This unit provides a platform from which students can think about challenges associated with the development of rural communities in the United States as well as current concerns about global climate change. Another example is the EEI high school biology unit that looks at genetic engineering from the perspective of California's Environmental Principle V, which concerns decision-making processes related to resources and natural systems, and helps students comprehend how political processes are directly related to key health and economic decisions about current-day concerns like the use of genetically modified organisms.

Using big environmental ideas as context gives students the chance to realize that similar situations produce similar results—whether locally or on the other side of the globe; in a rain forest ecosystem or the Sahara Desert; during the time of Paleolithic peoples or the nineteenth or twenty-first century. (The development of the EEI unit about Britain's Industrial Revolution is discussed in depth in the chapters that follow.)

Distinguishing among the concepts of "environmental context," "locations," and "systems" is important in developing an EBE program. A focus on environmental contexts is central to the process of creating EBE instructional programs and materials.

CHARACTERISTICS OF EFFECTIVE CONTEXTS

Selecting an environmental context in which to focus an EBE program requires investigating the available choices and making a series of decisions based on the specific needs of the individual program. Some of these choices are determined by the scale of the program and how it is going to be implemented. Programs at individual schools can realistically consider either of the two types of contexts, because they are likely to have access to a wide variety of local environment and community settings. Armuchee Elementary School in Rome, Georgia, for example, focused its program on the Armuchee Creek watershed because the program could be easily adapted to the local environment and community. Statewide or districtwide programs may be better served by focusing on an approach using big environmental ideas. The California EEI used the big environmental idea strategy because the curriculum had to be applicable to schools across the state, regardless of their local environments.

Whether the choice is to focus on a local environment or a big idea, selecting an instructionally effective context requires consideration of six important factors: (1) opportunities to connect to a wide variety of academic content standards; (2) efficacy toward developing students' understanding of systems (and associated higher-level thinking skills); (3) the diversity of opportunities for undertaking learning experiences that students will see as relevant to their daily lives; (4) developmental and age appropriateness for participants; (5) the potential for implementing environmental service-learning activities; and (6) a number of practical factors.

Connections to Standards

Relevance to standards-based instruction is one of the most important factors to take into account in choosing a context. The key consideration is the potential a particular environmental context offers for effectively teaching the adopted standards—Common Core State Standards for English language arts and math- ematics, as well as for science and history/social science standards. Whether based on a local environment or a big environmental idea, contexts that contain diverse natural and human social system components, processes, cycles, and interactions usually offer the most abundant opportunities for standards-based teaching.

With a local environmental context, it is much easier to teach multiple stan- dards in several disciplines in a school that contains or has ready access to some natural systems than in a school that has no such access. A school that has a local environment that includes a garden, pond, nearby park, or campus offers a broader range of possibilities for context-based lessons related to standards involving water quality and supply, pollination, decomposition, foods grown and eaten by early settlers, the effects of students' activities on the health of the pond, etc., and thereby offers greater opportunities for graphing and charting water data, writing persua- sive essays, and many other activities required by the standards. Teachers at a school that has limited access to natural systems can teach these same standards, but they face much greater challenges in engaging students and helping them learn about the interconnections among disciplines.

Environment-based education programs that choose to use big environmental ideas as their context must pay particular attention to the scope of those ideas. Big ideas that encompass diverse natural and human social system components, processes, cycles, and interactions provide numerous opportunities for teaching content standards. Environmental topics such as air pollution, which by their very nature are much more narrow, are relevant to teaching very few standards, and in and of themselves do not lead to a broad understanding of the connections between natural and human social systems. This also limits the number of standards that can be effectively taught.

Opportunities to Study Systems

The potential for developing students' understanding of systems and their use of systems thinking represents the second essential element involved in choosing an environmental context, whether local or based on a big idea. The context chosen must encompass the living and nonliving components of natural and human social systems, as well as the relevant cycles, processes, and systems interactions. A river, for example, while a wondrous place to learn about water, plants, insects, fish, birds, and other living things, cannot be viewed as a local environmental context for an EBE program unless it is connected with the processes, cycles, and the human social systems that might influence or be influenced by it.

If, for example, a team of teachers is planning to use a river as the environmental context for an EBE unit, examining the living and nonliving components of the natural elements of the river is just the starting point. "Knowing" about rivers means much more than simply studying and being able to describe the parts of the river system: water, plants, insects, fish, and birds. It also requires more than just learning about key river cycles and processes: the water cycle; erosion and how rivers transport soil to new areas; and the reshaping of rivers over time. A comprehensive understanding of a fully functioning river system must encompass knowledge of the river's components, processes, and cycles, along with the components, processes, and cycles associated with the human social systems related to it and how these two systems interact with each other. Figures 6.2, 6.3, and 6.4 present examples of how a river can be viewed, depending on whether it is thought of as a location, a natural system, or an environmental context.

FIGURE 6.2

A river as a location

Perspective	System components	Examples
River as a location	Natural system components	Water, soil, plants, insects, fish, birds, other living things, etc.

FIGURE 6.3

A river as a natural system

Perspective	System components	Examples
River ecosystem as a natural system	→ Natural system components	Water, soil, plants, insects, fish, birds, other living things, etc.
	→ Natural system processes and cycles	Water cycle, plant and animal reproduction, decomposition, soil erosion, etc.

FIGURE 6.4

A river as an environmental context

Perspective	System components	Examples
River as an environmental context	Natural system components	• Water, soil, plants, insects, fish, birds, other living things, etc.
	Natural system processes and cycles	• Plant and animal reproduction, decomposition, soil erosion, water cycle, nitrogen cycle, etc.
	Human social system components	• Laws, businesses, students and teachers, elected officials, telephones, Internet, cars and buses, highways and bridges, etc.
	Human social system processes and cycles	• Legal, economic, and educational systems, political systems, communications, transportation, recreational systems, etc. • Water runoff from transportation corridors like streets, highways, and parking lots • Use of fertilizers and pesticides on farms that enter the air and water systems • Smokestack emissions from factories
	Interactions between natural and human social systems	• Governmental decisions about the management of riparian habitats • Regulations on fisheries, including policies such as catch limits • Recreational uses like boating and fishing

Developing a context that includes the interactions between natural and human social systems allows for powerful learning experiences. Analyzing the connections and interdependencies among the different components, processes, and cycles within natural and human social systems and then evaluating them from the perspective of the interactions among these systems gives students an opportunity to develop higher-level thinking skills as they synthesize new information and apply these understandings to new situations.

Relevance to Students

Engaging students in learning what they perceive as relevant can be a complex task, because their interests vary tremendously, depending on their experiences and knowledge. Equally important, their interests change as they mature because of their increasing ability to perceive, understand, and relate to larger and more complex situations, concepts, and issues. Third-grade students might, for example, be interested in how hot playgrounds become without shade trees. Sixth-grade students, with their growing knowledge of their surroundings, usually begin to notice things from a larger perspective, such as school buses emitting "stinky" gases that they must walk through as they go into school. And by the time they reach high school, biology students' thinking and worldviews are much more developed, and they may become concerned if a nearby wooded area is threatened by the construction of a big shopping mall, and may want to know about the possible effects on native or endangered species.

There are also many ways to keep instruction interesting and relevant if big environmental ideas are used as the context. The California EEI curriculum accomplished this through its use of "California Connections" stories. Throughout the curriculum, stories like the one about the San Joaquin River Delta in a sixth-grade history/social science unit were developed to provide students with a personally relevant context within which to learn about rivers and deltas in ancient India and China, and at the same time, to relate this to Environmental Principle III (see appendix A at the back of the book).

Developmental Appropriateness for Students

The choice of context should also be evaluated from the perspective of whether it is developmentally appropriate for the age and maturity of the students. For kindergartners, thinking about global environmental topics is typically outside their realm of experience and ability to comprehend, and in most cases, it will not engage them intellectually or emotionally. Simply put, most cannot readily relate to information covering such large-scale issues. Most high school students, on the other hand, will be able to comprehend and logically evaluate and consider environmental problems at a global scale. Thinking about the choice of context in this way helps to ensure that learning will build on the students' existing knowledge and help them construct their own understandings of the new information they are learning, and thereby making it more meaningful to them.

Schoolwide EBE programs often select a local context and adapt it for each grade level so that learning is developmentally appropriate. At the Seven Generations

Charter School, from a large-scale perspective, the teachers are all using the environment, community, and surroundings of Emmaus, Pennsylvania, as the context. What varies at the different grade levels is the complexity and geographic extent of that context. At first grade, for example, one of the environmental contexts the students investigate involves the interactions of humans, animals, and weather with ponds in the local community. At sixth grade, the students are investigating a similar context, but at a much larger scale, studying the interactions among natural and human social systems in the Delaware River watershed as a whole. At each grade, the specific contexts have been selected to be age appropriate so that they can be readily connected to instruction in the relevant, grade-level academic content standards.

Seven Generations Charter School Local Environmental Contexts—Examples by Grade Level

Kindergarten: interdependence of plants, animals, and humans on the campus and in the local community.

First grade: interactions among humans, plants, animals, and the Emmaus community pond.

Second grade: interactions and effects of the Emmaus community on local forests.

Third grade: benefits the community gets from the local watershed, and how humans positively and negatively affect it.

Fourth grade: relationships among community resources, economic development, and Eastern forest ecosystems in the local area.

Fifth grade: interrelationships between local populations of pollinators (including bees), agricultural production, and other members of the community.

Sixth grade: interactions among natural components and processes of the creek near the school, the Delaware River watershed, and the school campus and local community.

Seventh grade: interactions between agriculture and the natural and human social systems of the Lehigh Valley.

Note: Each of the grades is studying several different contexts throughout the year, all relating to the local community.

Opportunities for Service Learning

Ideally, all EBE programs based on a local environmental context can provide students with opportunities to get actively involved in their communities by developing and implementing service-learning projects. Where this is possible, it may be a deciding factor in choosing among different environmental contexts and investigations. Once a context has been chosen, it is important to evaluate the realistic potential for students to implement different service-learning projects and to help them identify projects that are age appropriate and fit with their skills and abilities.

Sometimes students come up with ideas for their environmental service-learning work that they cannot realistically implement. For example, it is more likely that the third-grade students mentioned earlier who were concerned about the lack of shade on their playground could succeed with a tree-planting project than that they could undertake the construction of a shelter. Similarly, the sixth graders concerned about the exhaust from the school buses might successfully conduct an air pollution study and develop a plan to decrease the number of idling cars and buses in the driveway, but they would be much less likely to get their school district to replace the buses with an electrically powered fleet. While students will most certainly learn something even if their projects are not successful, they will have a greater sense of accomplishment and be more likely to get involved in other projects if they succeed in making an observable change in their community.

Service-learning projects provide students with the chance to develop many of the skills they need for college and/or careers and to become active members of a democratic society. For this reason, educators should look for opportunities for students to examine timely, real-world issues and to develop problem-solving, communication, and other twenty-first-century skills.

Practicalities

Practicalities make up the final factor to take into consideration in selecting a local environmental context for an EBE program. Time and funding are among the most significant things to consider in choosing among different contexts. For example, selecting a setting that is distant from the school can result in significant travel time and transportation costs. Some schools we have worked with have gotten around this challenge by involving parents and local businesses in transporting students to

Powerful Service Learning

Sixth-grade students at Huntingdon Area Middle School in Huntingdon, Pennsylvania, investigated Muddy Run, the creek that runs behind the school, and discovered a serious water pollution problem—the number of fecal coliform bacteria exceeded levels acceptable under state and federal water quality regulations. They brought this important problem to the attention of their borough council, and as a direct result of the students' work, the community received a grant of $250,000 from the Pennsylvania Infrastructure Investment Authority to repair the city's sewer lines—a system that was cracked and leaking into the community's creeks and streams. Equally important, for several years afterward, students continued scientific research projects on the organisms living in the creek as well as their efforts to actively engage members of their community in monitoring this problem. Studying this local environmental topic—something that was personally important to the students and had important health benefits for the community—was an experience that the students will never forget. These young people had the chance to discover firsthand how local democracy works.

field study sites. At one school, agents from a local real estate office transported the students to their study areas. This is not usually feasible, because of concerns over safety and liability. Similarly, it is impractical to select a context and investigation that require expensive equipment, books, or other resources to complete the work. Very few schools, for example, have access to the equipment necessary to test the chemistry of air or water pollutants. In some cases, however, this limitation can be overcome if sufficient time is invested in bringing diverse community partners into the program. For example, the students at Huntingdon Area Middle School were able to have water samples they collected from the creek tested by cooperating faculty and students at Juniata College, just down the road from the school—assistance that was ultimately critical to the credibility of their project.

Other practical matters to evaluate in deciding among various environmental contexts include: the level of technical expertise required to guide both students and teachers; the availability of volunteers, if needed, to support any investigations in the community; and safety concerns.

Key Considerations in Choosing an Environmental Context

The most successful choices among environmental contexts have certain characteristics. They:

- are applicable and beneficial to standards-based instruction in English language arts and mathematics, as well as in science, history/social science, and others, and have meaningful context for different grade levels;
- encompass living and nonliving components of natural and human social systems, as well as their interactions and the processes and cycles within and between these systems;
- support the development of higher-level thinking by allowing students to analyze real-world problems, apply knowledge and skills, synthesize information, formulate solutions, and use what they learn in new situations, and at the same time to build understanding of the interconnections among local, regional, and global concerns;
- offer opportunities to examine real-world problems and their connections to political, legal, economic, sociocultural, and natural systems, preparing students to participate in decision making in our civil society;
- deal with subject matter that is interesting and relevant to students, making their education understandable and meaningful;
- actively engage students in their communities so that they perceive how what they are learning connects to their daily lives, college choices, future jobs, careers, and service-learning experiences;
- are developmentally appropriate for the students in the program;
- provide students with opportunities to design and engage in environmental service-learning projects that they can successfully complete;
- are practical to implement in terms of such factors as time, distance from the school and availability of transportation, funding for materials, and technical support.

DEVELOPING A LOCAL CONTEXT

Teachers, ideally with the involvement of students, start the process of choosing an environmental context by exploring the area in and around the school. During and after their exploration, students and teachers create a community map that identifies all of the "things" they see around them—trees, garden area, playground, asphalt, driveways, streets, parks, houses, apartment buildings, schools, offices, businesses—whatever they see in the area.

Next, they add a more complex layer of information to their map: the natural and human social systems they observed. The natural systems include everything from a wooded area to a wetland to a farmer's field. The components of the human social systems they include on their maps should encompass transportation systems (streets, roads, highways, train tracks, and bus stations), communications and electrical systems (telephone lines, cell phone towers, power lines, transformers, and power plants), political and legal systems (government buildings, courthouses, police stations, and community meeting halls), etc.

As would be expected, more mature students can usually generate more accurate maps. Older students are also likely to include more details about the human social systems they observe, such as different types of businesses (e.g., retail establishments vs. auto repair shops and industrial facilities). These details can be important as students work to develop a deeper understanding of the interactions between specific systems, like business and industry, and local natural systems. Regardless of the sophistication of the maps, at all grade levels community mapping gets students involved in the process of identifying a local environmental context and at the same time becoming familiar with the school, the campus, the community, and their surroundings.

Student Involvement in Context Choice

Once the community mapping process has been completed, teachers and students can begin to analyze the local natural and human social systems and, from there, identify an appropriate local environmental context. Students at any grade level can be involved in the process of choosing the context, as long as they have guidance from their teachers.

In many cases, teachers take the lead by making the big picture choice of an environmental context. This is important, since the teachers have the responsibility for connecting their standards-based instruction with the context. They also need to ensure that the choice of context takes into account community-based investigations that are realistic in terms of considerations like distance and the availability of transportation, classroom management, and the accessibility of technical expertise.

Working with schools and districts implementing the Environment as an Integrating Context for learning (EIC) Model has shown us that the process of selecting an environmental context changes as teachers become more comfortable with implementing the program. At the earliest stages, they usually want to have full control over the context selection process. This is often appropriate, since they are still getting comfortable with connecting standards-based instruction within an environmental context. At the same time, students are discovering what it means to be learning through an environmental context. As teachers and students gain experience with EBE programs, the students can be given a greater role in choosing the local environmental context and the investigations they want to undertake.

The Texas Strands program at Brundrett Middle School in Port Aransas provides a good example of the process of gradually involving students in choices about context. The teachers made the initial decision to focus their EIC Model unit on the wetland habitats that dominate Mustang Island—an area that is within walking distance of the school and therefore a logistically manageable study area. When it came time to narrow the context focus, the teachers engaged their sixth- and seventh-grade students in the decision, and the students decided to focus the school's community-based investigations on the salt marshes at the Mustang Beach Airport and on other parts of the island.

Jackson Elementary School Chooses a Context

The fifth-grade "Water Works" unit developed by two teachers at Jackson Elementary School in Altadena, California, provides a useful example of how the community mapping process worked, and how the teachers and students further focused their context by identifying key components, processes, cycles, and interactions among the local natural and human social systems.[2] Since many fifth-grade science standards were related to the sources and use of water, the water cycle, and decreasing water use, the teachers decided that they wanted to work with their students to develop an EBE unit focused on water in their local community.

The development of the unit got underway with students and teachers walking around the campus and the school's neighborhood, finding some satellite photos on the Internet, and locating some maps of the area. The teachers put a special emphasis on gathering information about water use and availability in the school, on the campus, and in the neighborhood. After a few days, they had all the information they needed, so the two classes created classroom maps that brought together everything they had discovered during their exploration. Figure 6.5 presents a facsimile of the map developed by Jackson's fifth-grade students and teachers. (Later

FIGURE 6.5

Facsimile of Jackson Elementary School's fifth-grade community map

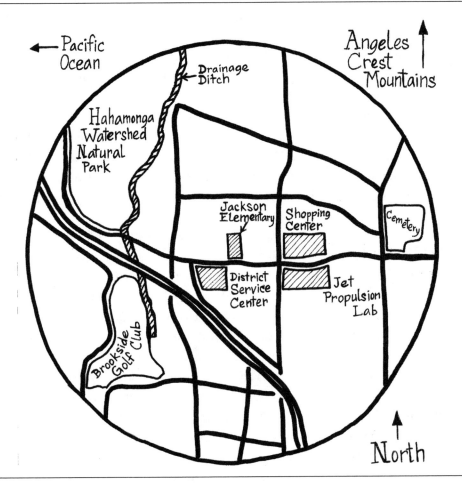

in the year, as the students' exploration expanded further into the community, they incorporated elements relating to the flow of water from campus to the local water treatment plant and beyond to their maps.)

With their exploration of the community and the initial versions of their maps completed, students and teachers began the process of organizing what they had seen into two groups of components—natural and human social. They built onto this relatively simple, piecemeal view of the world by listing as many of the cycles, processes, and interactions that they were already aware of in these natural and human social systems.

Components, Cycles, and Processes Identified by Jackson's Fifth Graders

Natural Systems

- Natural water systems and sources, including the water cycle, groundwater, rivers, and runoff from Sierra Nevada mountain range
- "Ecosystems," as well as natural and human-made habitats, including: "green" areas like the campus lawns, gardens, and playing fields near the school; local parks, habitats, and areas around students' homes

Human Social Systems

- Education system, including: students, teachers, administrators, families, and community members
- Water transport system, including: reservoirs, water supply lines, storm drains, sewage lines, and water treatment facilities
- Economic system, including: school budgets, community budgets, and taxes
- Transportation system, including: sidewalks, roads, cars, and buses

The next stage of the process was to identify the interactions among these natural and human social system components and processes.

Natural and Human Social Systems' Interactions Identified by Jackson's Fifth Graders

- Water and soil from campus areas flooding into the school building, making a classroom unsafe and increasing costs to the school and community
- Water from the natural water system being used by the school and community
- Soil loss due to erosion, and damage to the vegetated areas on campus, including lawns, gardens, and playing fields
- Excess water from the water system affecting the "green" areas on campus and entering the storm drains, increasing costs to the school and community

Finally, to complete the development of their environmental context, the students identified how the interactions would affect each of the systems.

Examples of Effects of Systems' Interactions Identified by Jackson's Fifth Graders

How the Interactions Affect the Natural Systems
- Water used by the school and community affects the quantity of water in rivers and lakes;
- Soil is removed from the natural systems on campus and in the community;
- Campus vegetation is damaged by excess water;
- Ocean pollution results from soil and other materials flowing downstream through storm drains.

How the Interactions Affect the Human Social Systems
- Water damaged the school building by flooding the fourth-grade classroom;
- Issues with water and soil increased school health and safety concerns;
- Greater water costs for the school and community resulted from excess soil and other materials clogging the storm drains and requiring costly cleanup work.

It was these components, processes, cycles, and interactions that became the environmental context for the "Water Works" unit and the overarching framework for their fifth-grade EBE program.

As Ms. Peters and Mr. Silverio developed this unit, they took into account all the key factors for developing an effective local environmental context. They started by focusing on the standards they had set as priorities, including English language arts, math, science, and history/social science. Next, based on the students' interest, they zeroed in on water waste issues on campus. These enthusiastic teachers also used the unit to engage their students in a variety of field experiences and hands-on activities—like collecting water usage data on campus and at home—to strengthen higher-level thinking skills and develop greater understanding of systems and systems interactions.

BIG IDEAS AS A CONTEXT

As discussed earlier, big environmental ideas can serve well as the context for EBE programs. Many of the considerations related to choosing a context based on big environmental ideas are similar to those involved in selecting an appropriate local environmental context. The differences between the two mainly stem from the fact that EBE programs based on big environmental ideas cannot always take into account local and community-based environmental concerns.

There is no widely accepted set of big environmental ideas to use as the basis for organizing EBE programs. When we began California's process of identifying environmental principles, we undertook a global search for existing materials. Not finding any comprehensive sets of big environmental ideas, California developed its own set of Environmental Principles and Concepts, as described briefly in chapter 4. This process took many months, because most of the education programs developed by environmental agencies and organizations have focused on topical content dealing with the effects of human activities on a particular species or ecosystem rather than on generalizable, overarching principles. The difficulties of developing California's principles continued, even when many experts were brought together to support the process, because almost all of these individuals had such specialized, in-depth knowledge about their content area that it was difficult for them to step back and consider the environment from the perspective of big, organizing ideas.

Britain Solves a Problem

California's Education and the Environment Initiative (EEI) uses a somewhat different approach to establish the environmental contexts that are the focus for instruction in each of its units. It is a three-pronged approach that combines one or more science and history/social science content standards with one or more of California's Principles, and then ties that to a relevant "California Connections" story that discusses a present-day environmental topic in the state. The big environmental idea context for "Britain Solves a Problem and Creates the Industrial Revolution," one of the EEI's tenth-grade world history units, is California's Principle I: "The continuation and health of individual human lives and of human communities and societies depend on the health of the natural systems that provide essential goods and ecosystem services." The California Connections story for "Britain Solves a Problem" focuses on a story that most California high school students are familiar with—innovations in electronics, the computer revolution, Silicon Valley—and how these changes influenced Santa Clara County's land-use patterns and agricultural systems. These modern-day topics were selected to help students frame their understanding of the world history standards in a way that they can readily understand. They also allow students to perceive the relevance of these historical events, processes, and outcomes to their daily lives. The appendix at the end of this chapter summarizes the context and the natural and human social systems that are the focus of the tenth-grade world history unit "Britain Solves a Problem and Creates the Industrial Revolution."[3] It also provides examples of the interactions among the systems and the effects of those interactions on each of the systems.

The examples of the environmental contexts at Jackson Elementary School and in the EEI— based on an EIC Model unit and California's EEI curriculum, respectively—demonstrate the two major approaches to using environmental contexts to teach state standards to a high level of proficiency. At the same time, they also exemplify a few of the many opportunities that environmental contexts can provide for students to simultaneously learn about natural systems and human social systems and their interactions.

After the environmental context is determined, EBE planners can begin the process of selecting standards and writing learning objectives and instructional materials, the subject of the chapters that follow.

APPENDIX (CHAPTER 6): INSTRUCTIONAL FRAMEWORK OF CALIFORNIA'S TENTH-GRADE EEI UNIT "BRITAIN SOLVES A PROBLEM AND CREATES THE INDUSTRIAL REVOLUTION"

Vision and Mission

"Increasing Environmental Literacy for K–12 Students . . . Because the Future Is in Their Hands."[4]

Environmental Context

"Big" Idea(s) as the Environmental Context

California Environmental Principle I: The continuation and health of individual human lives and of human communities and societies depend on the health of the natural systems that provide essential goods and ecosystem services.

Concept a: Students need to know that the goods produced by natural systems are essential to human life and to the functioning of our economies and cultures.

Concept b: Students need to know that the ecosystem services provided by natural systems are essential to human life and to the functioning of our economies and cultures.

"California Connections" story

New Challenges, New Opportunities, New Technology

Natural system(s)

England's forests, rivers, water supplies, and agricultural lands

Human social system(s)

England's economic system (before, during, and after the Industrial Revolution), transportation system, laborers, population growth, inventions of the later eighteenth and early nineteenth centuries, human health

Interactions among systems (selected examples)

- Transformation to industrial economy, and changes to forested lands, water quality, air quality, agricultural land, and availability of natural resources
- Changes in demands for natural resources resulting from growth of the human population

Potential effects of the interactions on the natural systems (selected examples)

- Water pollution influencing health of river ecosystems
- Air pollution influencing health of forest ecosystems
- Changing demand for energy resources (wood and coal) in order to fuel factories, resulting in decreased forest habitats and increased disturbance to areas with coal supplies

Potential effects of the interactions on the human social systems (selected examples)

- Additional opportunities for jobs at factories, coal mines, and in harvesting of wood
- Movement of large numbers of people from rural areas to urban areas
- Increased health issues resulting from greater consumption of fuels harvested from natural systems

Standards

History/Social Science Standards

10.3.1. Students "analyze why England was the first country to industrialize."

10.3.5. Students "understand the connections among natural resources, entrepreneurship, labor, and capital in an industrial economy."[5]

Learning Objectives

Students will:

- Recognize natural systems and the resources they provide (ecosystem goods and ecosystem services) as the basic capital for the development of an industrial economy.
- Provide examples of the major connections between natural systems and resources, and entrepreneurship, labor, and capital in industrial economies (for example, the labor necessary to extract, harvest, transport, and produce ecosystem goods and ecosystem services for human communities).
- Describe how increased demands provided an economic opportunity for the English to improve the methods they used to extract, harvest, transport, and produce goods from the natural resources that were available.
- Recognize that the growth in human populations and human communities in England placed greater demands on natural systems.

CHAPTER SEVEN

Connecting Standards to an Environmental Context

The 2010 release of the Common Core State Standards for English language arts and literacy in history, social studies, science, and technical subjects, and for mathematics, and their formal adoption by forty-five states and three territories (as of 2013) are significantly changing the landscape of education. Reinvigorating standards-based instruction by focusing on relevance to the real world was one of the major goals driving the Common Core State Standards Initiative. As described in the initiative's mission statement: "The standards are designed to be robust and relevant to the real world, reflecting the knowledge and skills that our young people need for success in college and careers."[1]

The new English language arts/literacy standards, for example, speak to the need for an interdisciplinary approach that can develop students' "ability to gather, comprehend, evaluate, synthesize, and report on information and ideas, [and] to conduct original research in order to answer questions or solve problems."[2] And the mathematics standards point to the need for "proficient students [who] can apply the mathematics they know to solve problems arising in everyday life, society, and the workplace," ". . . [write] an addition equation to describe a situation . . . and apply proportional reasoning to . . . analyze a problem in their community."[3]

The English language arts/literacy standards, in particular, speak repeatedly to the importance of developing proficiency with these skills in concert with science, history/social science, technical subjects, and the arts. Environment-based education (EBE) is an important strategy for achieving this goal, since it can provide an effective context for integrating instruction across these disciplines, thereby offering students diverse opportunities to strengthen what the Common Core calls the "integration of knowledge and ideas."

The status of science standards is somewhat different from that of the disciplines included in the Common Core State Standards. The National Research Council, National Science Teachers Association, American Association for the Advancement of Science, and Achieve, Inc., led the effort to compile "Next Generation Science Standards" based on the National Research Council's 2011 *A Framework for K–12 Science Education*. Representatives from twenty-six state departments of education were actively involved in this process, making it likely that these standards will be widely adopted across the nation.

Across all grades, these proposed national science standards include many opportunities for study that encompass the components, processes, and cycles within natural systems and how they are affected by human activities. This starts at the kindergarten level with disciplinary core ideas such as "Human Impacts on Earth Systems: Things that people do to live comfortably can affect the world around them. But they can make choices that reduce their impact on land, water, and other living things."[4] This focus continues into the high school grades, where students are to "evaluate or refine a technological solution that reduces impacts of human activities on natural systems."[5]

The state of affairs of standards for the various disciplines encompassed by history and the social sciences differs greatly from both the Common Core State Standards and the new science standards. There is not a single widely accepted compilation of standards covering this content. Over the past twenty years, a number of organizations have developed diverse sets of proposed national standards covering these disciplines, including: the National Council for the Social Studies, which in 2010 released *National Curriculum Standards for Social Studies: A Framework for Teaching, Learning, and Assessment*; the Council for Economic Education, which in 2010 published an updated version of *Voluntary National Content Standards*; the Center for Civic Education, which in 2007 updated its *National Standards for Civics and Government*; the National Center for History in the Schools (under the guidance of the National Council for History Standards), which in 1996 released the *National Standards for History: Basic Edition*; and the National Geographic Society, with its 1994 publication of *National Geography Standards*.

In history/social science, like all the other disciplines, each state has the final responsibility for developing and adopting its own standards. However, with the diverse input from all the different professional and nongovernmental organizations, and no single, national compilation of standards for history/social science content, there is a much greater diversity among states' standards in these content areas.

IDENTIFYING CURRICULUM STANDARDS

Once an environmental context is selected, the next step in creating an environment-based education (EBE) program or unit is identifying which standards are most effectively taught through the chosen context and by means of the type of interdisciplinary investigations and service-learning projects that EBE programs typically include. The important point is that EBE programs are designed to work with curriculum standards from any national, state, or district source. This chapter will show how educators can systematically evaluate and select standards best suited for the context they have chosen and for students' simultaneously learning about natural and human social systems. This step is critical for ensuring that students will be learning key academic content knowledge at the same time they are learning about the local environment or big environmental ideas. This step is also what distinguishes EBE from traditional environmental education.

Environment-based education programs are especially useful for helping students become proficient in the Common Core State Standards, as well as in science and history/social science, because they share many of the same higher-order learning goals. But they can also be used as the basis for combining instruction from other disciplines, like technology, art, and music. The process of designing an EBE program for a state, school district, or individual school focuses on the adopted academic standards, as EBE programs are flexible enough to be readily adapted to the diversity of standards found across the country.

A rich range of opportunities for using the environment as a context is readily revealed by analyzing: already adopted Common Core State Standards; the final version of the Next Generation Science Standards; and a cross-section of state level history/social science standards. Much of the content and many of the skills that are represented by the standards in each of these disciplines can be easily incorporated into EBE programs, as demonstrated by the examples below. Standards identified in these examples have been built into the EBE programs described throughout this book:

- **English language arts standards:** Common Core State Standards that, among many other things, require students to develop and use research and media skills to gather, comprehend, evaluate, synthesize, and report on information and ideas; conduct original research in order to answer questions or solve problems; and analyze and create a wide range of print and nonprint texts in the form of both old and new media.

- **Mathematics standards:** Common Core State Standards that, among many other things, have students: develop and use these skills in the context of solving real-world and mathematical problems; summarize, represent, and interpret numerical data in relation to their context; draw inferences about populations from collections of data; model with mathematics; and develop, use, and evaluate probability models.

- **Science standards:** As represented in the Next Generation Science Standards, which, among many other things, have students: develop their science and engineering practices, such as planning and carrying out investigations, analyzing and interpreting data, and communicating information; learn about the components of natural systems, including water, soil, plants, insects, fish, birds, and other living things; investigate weather, climate systems, and patterns of climate change, as well as Earth systems and their interactions; and delve deeply into broad issues, such as human impacts, sustainability, and links among engineering, technology, science, and society.

- **History/social science standards:** Standards from a cross-section of states that, among many other things, have students: learn about legal systems, economic systems, educational systems, political systems, communications systems, and transportation systems; examine the influence on human populations of physical systems such as climate, weather and seasons, and natural resources; discover that resources are limited, and therefore people cannot have all the goods and services they want; evaluate different methods of allocating goods and services by comparing the benefits to the costs of each allocation method.

Environment-based education programs are intended to function as an integral part of a school's, district's, or state's standards-based academic instruction. Once they have identified their environmental context, educators are ready to begin identifying the standards that will be the focus of the instructional units within their program. They analyze academic content across the disciplines, choosing the standards from each subject that they believe can be taught effectively in their particular environmental context. As they do this analysis, educators choose two types of standards: discipline-specific academic standards and standards related to teaching about systems.

Discipline-Specific Standards

Selection of the content standards that will be incorporated into an EBE program requires careful consideration and purposeful decision making. In general, standards from any of the disciplines can be grouped into one of three categories: those that can be fully taught in an environmental context; others for which learning can be enhanced through context-based instruction; and a final group that is difficult or inappropriate to teach through an environment-based approach.

Most EBE programs focus on the first category of standards—those that can be taught to proficiency in an environmental context. For example: Pennsylvania's sixth-grade geography standard 7.4.6.A.—"Describe and explain the effects of the physical systems on people within regions"—is taught at Seven Generations through an investigation focused on water use in the Delaware River watershed; and California's twelfth-grade EEI unit economics standard 12.2.2.—"Discuss the effects of changes in supply and/or demand on the relative scarcity, price, and quantity of particular products"—is taught through analysis of global fisheries resources. Many of these programs, however, also include instruction related to standards teachers believe will be enhanced if taught in a context that is meaningful to the students.

Having students read, write, and speak about content that interests them can, as emphasized in the English language arts/literacy Common Core State Standards, benefit students' learning. All of California's kindergarten through third-grade Education and the Environment Initiative (EEI) units, for example, in addition to covering content from the science or history/social science standards, contain instructional activities intended to help students strengthen some of their basic English language arts skills. For example, third-grade students may work on Common

Core English language arts standards such as comprehension and collaboration skills by reading about a desert animal and an animal from a local ecosystem, and then discussing how each animal's survival depends on a healthy ecosystem. Or they may write informative/explanatory texts to examine a topic and present ideas by writing paragraphs that compare and contrast the costs and benefits incurred by local industries in the past and present.

The process of examining disciplinary standards and organizing them, at least preliminarily, into these three categories is an important first step in selecting the standards that will guide the development of an EBE program. Educators who are developing instruction based on the Environment as an Integrating Context for learning (EIC) Model undertake this process from the very beginning. Working in either grade-level and/or discipline-specific teams, they review all their standards and highlight those they think can be most effectively taught in their local context. The first stage of developing the EEI curriculum followed a very similar strategy— each science and history/social science standard for kindergarten through twelfth grade was reviewed and evaluated in terms of its potential connection to California's Environmental Principles and Concepts. This initial step helps EBE planners narrow the breadth and number of standards that are being considered. But it is just the beginning, not the end of the standards selection process, because it invariably leaves educators with large numbers of standards that they want to teach in an environmental context. In a school setting, teachers often identify over 100 standards across the disciplines; over 300 standards met the initial criteria that were used for the California EEI.

Since it is not usually feasible to develop a program that encompasses hundreds of standards, EBE planners take the next step of narrowing down their choices. They work their way through the remaining standards and rank them based on three key criteria—standards that:

- address abstract ideas or skills that have otherwise proven difficult for students;
- are more meaningful to students if taught in a real-world context;
- can be taught effectively when connected to instruction in other standards within the same or another discipline.

Teachers at Seven Generations Charter School used this process to select the standards they would use to create an EBE program focused on their local environment of Emmaus, Pennsylvania. Working in grade-level teams, all the teachers reviewed the Common Core State Standards as well as other state- and district-adopted standards. As they examined all of the standards for English language arts, mathematics, science, social studies, and technology, they asked themselves three key questions aligning with the criteria listed on the previous page:

- Which of these standards represent abstract content or concepts that would be more readily learned in an environmental context that can provide real-world examples and opportunities for hands-on activities?
- Which of these standards can be learned in a more meaningful way, so that students will be better able to apply their knowledge and skills to their daily lives?
- Which of these standards can be taught through an integrated interdisciplinary approach?

Once they had answered these questions, they asked each other one final question: "If taught using an EBE strategy, are students more likely to achieve proficiency with these standards and succeed academically?"

Having completed their analysis of the standards, and after assessing their students' progress up to that point in the school year, the Seven Generations teachers identified those standards that they believed could be most effectively taught through a curriculum centered on their local environment. The teachers also chose several standards with which their students had previously had difficulty—standards that when taught only through textbooks were conceptually abstract. They felt that a community-based investigation would make the content more tangible and meaningful for their students. Teachers at all grade levels followed this same process and identified their own sets of standards.

The fifth-grade team, for example, designed a unit called "Nature's Pollinators," which included a community-based investigation intended to help students succeed with standards in several disciplines. Examples of some of these standards include:

- **Math:** "Convert among different-sized standard measurement units within a given measurement system, and use these conversions in solving multi-step, real world problems"; and "Make a line plot to display a data set of measurements in fractions of a unit. Use operations on fractions for this grade to solve problems involving information presented in line plots." (Common Core State Standards 5.MD.1 and 5.MD.2)

- **English language arts:** "Compare and contrast the overall structure (e.g., chronology, comparison, cause/effect, problem/solution) of events, ideas, concepts, or information in two or more texts." (ELA Common Core State Standard 5.R.IT.5)
- **Science and technology and engineering education:** "Compare and contrast the similarities and differences in life cycles of different organisms." (Pennsylvania Standard 3.1.5.A3)

The Seven Generations teachers decided to focus their work on a topic that the students had expressed interest in during their exploration of the community—the decline of the local bee population, a problem that is facing many agricultural communities around the United States. Using the standards above, the fifth-grade teachers designed a series of lessons that would allow them to teach the standards they had identified during their initial analysis, as well as several other Pennsylvania social studies, and environment and ecology standards.

As they built a beehive, measured for their garden plots, and monitored the growth of plants in the garden, the students developed their measuring skills and learned about standard measurement units. Their investigation into the declining local bee population engaged students in collecting data about plants, animals, gardening, and crop production. They made line plots of the information they had gathered to help them analyze the situation and draw conclusions based on their data. These fifth-grade students then made firsthand observations of bee and butterfly boxes and watched videos as the basis for creating their own diagrams of bee and butterfly life cycles, which they then compared and contrasted to examine the similarities and differences between them.

Having worked through this initial sorting process, EBE program planners will have identified a subset of standards in each discipline that they will consider teaching through their instructional unit(s). Most program designers focus on groups of standards in different disciplines that they plan to teach during a particular time frame, such as over a specific multiweek period, semester, or even a season. As they review and analyze their standards, they must also consider which can most effectively be taught in connection with the chosen environmental context.

Systems-Related Standards

As the next step in the process, an EBE program design team revisits their initial subset of standards, this time with the purpose of identifying the standards that will help them connect their discipline-specific teaching to instruction about natural and human social systems. Standards directly related to these systems and their interactions are usually found within the history/social sciences, life sciences, and earth sciences. For example, within history/social sciences, standards relevant to developing an understanding of systems can often be found within content related to civilizations, trade, migration, colonization, wars throughout history, geography, and many other realms. In the sciences, as represented in the Next Generation Science Standards, studies of systems are important for students across grade levels, as systems and system models are included explicitly as one of the major "crosscutting concepts" within the sciences.

The "Nature's Pollinators" unit illustrates the next important step in the process of developing an EBE program: selecting the systems-related content (contained in standards or big environmental ideas) that will be the basis for teaching about natural systems and how they influence and are influenced by human social systems. In this case, the teachers chose Pennsylvania's science and technology and engineering education standard 3.4.5.D3: "Determine if the human use of a product or system creates positive or negative results." Throughout the "Nature's Pollinators" unit, students were engaged in analyzing the information they had collected and looking for evidence of interactions between the natural system—including bees, plants, and pollination—and human social systems such as gardening and farming, changing land use, and use of herbicides and pesticides. (See the box titled "Using Science to Investigate a Mystery" in chapter 3 for one teacher's thoughts about the students' success with this unit.)

The creation of California's EEI curriculum illustrates another approach—determining which standards can be taught to mastery in the context of each of the five environmental principles, and fifteen supporting concepts. Over several months of review, the state of California's five educational consultants completed an analysis of each of the state's approximately one thousand K–12 science and history/social science standards.

The review involved asking and answering a series of questions about each standard: "Will students' proficiency with this standard increase if it is taught in concert with one or more of the Environmental Principles and Concepts?" "Can the standard be more fully taught through a traditional textbook-based approach or in the context of natural and human social systems?" and, finally, as a result of teaching the standard through the EEI curriculum: "Will students further their understanding of the Environmental Principles and Concepts and be better able to apply these ideas to issues they might encounter?"

The consultants used the answers to these questions as the criteria for grouping the standards into two major categories: standards potentially relevant to the EEI and standards clearly inappropriate for inclusion in the curriculum. Since the goal of California's Department of Education and State Board of Education is for students to achieve "mastery" of the standards, dividing them into these groups was a relatively straightforward process—Newton's laws, atomic and molecular structure, and the moral and ethical principles of religion are just a few examples of standards that were easily placed into the "inappropriate for inclusion" category.

Through this sorting process, the EEI planning team identified promising connections to natural and human social systems among about 300 of California's science and history/social science standards. They then undertook a second level of review, looking for standards that could be effectively used to teach about one or more of the Environmental Principles and Concepts.

One eighth-grade standard exemplifies the two-step process. In reviewing California's history/social science standards, the consultants found one, 8.12.1., that deals directly with the interactions between systems: "Trace patterns of agricultural and industrial development as they relate to climate, use of natural resources, markets, and trade and locate such development on a map." They then analyzed how the standard could be connected with the ideas covered by the Environmental Principles and Concepts and identified strong connections with Environmental Principle II, Concept a: "[D]irect and indirect changes to natural systems due to the growth of human populations and their consumption rates influence the geographic extent, composition, biological diversity, and viability of natural systems," and Concept c: "[T]he expansion and operation of human communities influences the geographic extent, composition, biological diversity, and viability of natural systems."

The EEI consultants decided to propose this standard, and many others, to the Interagency Model Curriculum Planning Committee for inclusion in the EEI curriculum because of the direct connections the standard makes between human developments (agriculture, industry, markets and trade) and natural systems (climate, natural resources). This decision eventually led to the creation of an eighth-grade unit called "Agricultural and Industrial Development in the United States."

Over many more months, and in consultation with the planning committee, this process was used to narrow the selection of standards for the EEI curriculum even further. Ultimately, development of the EEI curriculum was focused on 101 of California's science and history/social science standards.[6]

IDENTIFYING INTERDISCIPLINARY CONNECTIONS

Having identified their discipline-specific and systems-related standards, EBE program planners must start to think about how they will connect these standards within their instructional units. In the EIC Model, educators organize units around "community-based investigations" in their local environmental context; programs focused on teaching big environmental ideas, like California's EEI, organize units around "investigations" related to the systems-based standards that the educators have chosen. For example, "River Systems and Ancient Peoples" is a sixth-grade EEI unit developed around California's history/social science standard 6.2.1: "Locate and describe the major river systems and discuss the physical settings that supported permanent settlement and early civilizations." These types of investigations are effective ways to teach discipline-specific and systems-related standards at the same time as students are learning about either their local environmental context or a big environmental idea.

These investigation-centered approaches offer diverse opportunities for creating instructional units that give students a wide range of interdisciplinary learning experiences. While an interdisciplinary approach to curriculum and teaching is not new, EBE programs offer a contemporary and exciting vehicle to inject more of this type of teaching and learning into schools. Educators have found that local environmental contexts and big environmental ideas offer exceptionally good opportunities for integrating instruction across the disciplines because studying and understanding them requires knowledge and skills drawn not only from science and history/social science, but from, among others: communications (English language arts/literacy); quantitative analysis (mathematics); systems analysis; use of advanced information, media, and technologies; collaborative problem solving; and civic involvement.

The fifth-grade "Water Works" unit developed by the teachers at Jackson Elementary School in Altadena, California, is a good example of how a community-based investigation was used to integrate instruction in English language arts, mathematics, science, and history/social science. To begin the unit, students conducted water studies on campus to gain content specified in state science standards—learning about the origin of the water used in their local community and identifying ways to decrease water use through both conservation and recycling. The teachers then used the students' research as the basis for reading and discussing how geography and climate influence water distribution. They connected this information to content outlined in history/social science standards through additional readings and discussions about how people in different nations adjust their behaviors and the locations of their villages according to available sources of water. Students learned math content by measuring the volume of water used in different locations and computing basic statistics, and making estimates related to the data they collected about campus water usage. Students then used their data as the basis for making plots, including histograms and bar graphs. To strengthen their English language arts skills, in addition to doing research on the Internet, students incorporated everything they had learned through their science, history/social science, and mathematical analyses into written reports and presentations. Table 7.1 summarizes some of the interdisciplinary connections that the teachers made as the students worked on their community-based investigations within "Water Works," a fifth-grade EIC Model unit.[7]

The process of developing interdisciplinary connections requires educators to think about the investigations that are going to drive their units—either community-based or focused on a big environmental idea. As they develop their units, they decide which standards in each discipline they will teach in the context of their investigations. For example, if their community-based investigation is focused on water usage, it is logical to teach the students science standards connected to water supply; math standards related to measuring, volume, data analysis, and statistics; English language arts standards targeted at reading technical documents, writing expository text, and presenting to groups; and history/social science standards that examine the function of local government. In this way, educators can

TABLE 7.1

Interdisciplinary connections in Jackson Elementary School's "Water Works" unit

English language arts & literacy Common Core State Standards (selected examples)[1]

- Draw on information from multiple print or digital sources, demonstrating the ability to locate an answer to a question quickly or to solve a problem efficiently.
- Write informative/explanatory texts to examine a topic and convey ideas and information clearly.
- Include multimedia components (e.g., graphics, sound) and visual displays in presentations when appropriate to enhance the development of main ideas or themes.

Mathematics Common Core State Standards (selected examples)[2]

- Make a line plot to display a data set of measurements in fractions of a unit. Use operations on fractions for this grade to solve problems involving information presented in line plots.
- Convert among different-sized standard measurement units within a given measurement system, and use these conversions in solving multi-step, real world problems.
- Relate volume to the operations of multiplication and addition and solve real world and mathematical problems involving volume.

Science standards (selected examples)[3]

- Students know that the amount of fresh water located in rivers, lakes, underground sources, and glaciers is limited and that its availability can be extended by recycling and decreasing the use of water.
- Students know the origin of the water used by their local communities.

History–social science standards (selected examples)[4]

- Describe how geography and climate influenced the way various nations lived and adjusted to the natural environment, including locations of villages, the distinct structures that they built, and how they obtained food, clothing, tools, and utensils.
- Discuss the experiences of settlers on the overland trails to the West (e.g., location of the routes; purpose of the journeys, the influence of the terrain, rivers, vegetation, and climate; life in the territories at the end of these trails).

1. National Governors Association Center for Best Practices and the Council of Chief State School Officers, *Common Core State Standards for English Language Arts & Literacy in History/Social Studies, Science, and Technical Subjects* (Washington, DC: National Governors Association Center for Best Practices and the Council of Chief State School Officers, 2012), 14, 20, 24. The author has converted the original California English Language Arts standards to the language of the comparable Common Core State Standards.

2. National Governors Association Center for Best Practices and the Council of Chief State School Officers, *Common Core State Standards for Mathematics* (Washington, DC: National Governors Association Center for Best Practices and the Council of Chief State School Officers, 2012), 37. The author has converted the original California Mathematics standards to the language of the comparable Common Core State Standards.

3. California Department of Education, *Science Content Standards for California Public Schools: Kindergarten Through Grade Twelve* (Sacramento: California Department of Education, 1998), 16.

4. California Department of Education, *History–Social Science Content Standards for California Public Schools* (Sacramento: California Department of Education, 1998), 16, 20.

simultaneously teach multiple content standards and help their students understand the interconnections among knowledge and skills from diverse disciplines. Figure 7.1 illustrates the point that this strategy requires that connections be made among standards as well as with the environmental context.

Teachers who implement what EBE programs call "integrated interdisciplinary instruction" (I.I.I.) discover that, by making interdisciplinary connections among the standards, some of their instructional practices are more effective for their students, and also help them make more efficient use of instructional planning and class time. Educators using an I.I.I. strategy observe and have reported several

FIGURE 7.1
Integrated-interdisiplinary connections

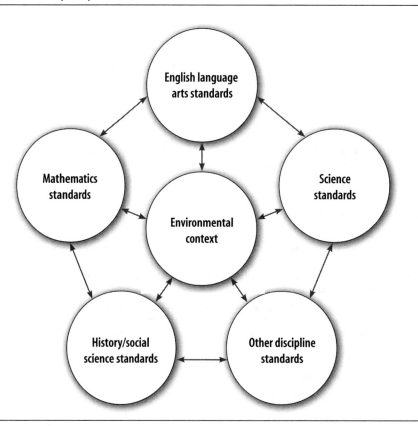

other benefits for their students and themselves. A majority of teachers report that, when learning through interdisciplinary instruction, their students perceive school as more meaningful and exhibit improved comprehension of difficult subjects. Teachers also say that I.I.I. provides them with more opportunities to reinforce each other's instruction.

Interdisciplinary EBE units also provide many opportunities for students to develop higher-level thinking skills, as demonstrated by another Seven Generations Charter School example. An interdisciplinary team of teachers—Pamela Kattner (science), Hilary Heffner (history/social studies), Souzan Boules (mathematics), and Joe Brisgone (English language arts)—recently developed a sixth-grade unit at the school that gives students the opportunity to study how seasonal cycles in the Tigris-Euphrates and Nile River valleys influenced the development of permanent settlements in the area by early civilizations. The teachers' I.I.I. strategy connects instruction called for by Pennsylvania's standards for math, English language arts, history/social science, and science, and allows them to simultaneously focus on content for several standards, including: historical settlement patterns and natural resource–based industries; the impact of watersheds and wetlands on people; the

Benefits of Making Interdisciplinary Connections

The experience of teachers in EBE programs indicates that integrated interdisciplinary connections can:

- develop students' higher-level thinking skills by allowing them to examine interdisciplinary connections;
- offer opportunities for students to work simultaneously on interrelated aspects of the same topics and apply what they are learning to several subject areas during multiple class periods;
- change students' perception of the importance of what they are learning to their daily lives;
- help students with content that might otherwise be difficult for them to comprehend;
- facilitate and reinforce instruction in each of the interconnected standards.

use of variables in real-world problem solving; and the gathering of relevant information from multiple print and digital resources. This unit strengthens students' higher-level thinking skills by challenging them to analyze the data they have gathered about the current use of the Delaware River and its watershed, apply it to their studies about how early civilizations used rivers, and to then evaluate and write about the long-term effects of the changes taking place near them.

Teaching teams in EBE programs use I.I.I. because it gives students opportunities to work simultaneously on interrelated aspects of the same topics and to apply what they are learning to several subject areas during multiple class periods. High school biology students, for example, study the global distribution of ecosystems that share similar climates and vegetation. At the same time, they can work with their history/social sciences teacher to examine how these ecosystems influenced colonization during the nineteenth century by looking at the goods and services that were being extracted, and the costs and effects of this use on the natural systems. Their mathematics and English language arts teachers can then use what they have learned in their other courses as the content for real-world application of Common Core State Standards focused on statistics and on writing arguments based on claims, counterclaims, reasons, and evidence.

The I.I.I. approach changes students' perception of some of the material they are studying, helping them recognize how it is meaningful to their daily lives. Most fourth-grade students, for example, are uncertain as to why they are learning about the legislative process. The students at Armuchee Elementary School in Rome, Georgia, however, saw firsthand the practical value of what they learned about how state laws are written and passed when they sought to get the green tree frog recognized as Georgia's official state amphibian by the legislature and governor. They made their case by combining their newly acquired scientific knowledge of the frogs with their English language arts communication skills. Their letters and meetings with the states' legislators made all the difference, and eventually they succeeded.

Some content is difficult for students to comprehend if it is taught in isolation. Mathematics learning, in particular, benefits from being combined with real-world projects. Students from Pennsylvania's Huntingdon Area Middle School had this opportunity after they collected scientific data about water quality, stream flow, stream habitat, and the invertebrates living on the stream bottom from four study sites at a nearby creek. This research gave students the opportunity to study a variety of mathematics standards in the context of their own real-world data. Working with their math teacher, they studied a range of Pennsylvania standards, including: "Analyz[ing] data and/or answer[ing] questions pertaining to data represented in frequency tables, circle graphs, double bar graphs, double line graphs or line plots"; and "Determin[ing]/calculat[ing] the mean, median, mode and/or range of displayed data."[8] Using their quantitative data, Huntingdon's students calculated a "Cumulative Stream Quality Index Assessment" for each site based on the number and diversity of stream invertebrates and on three levels of sensitivity to pollution. They discovered that some nutrient loading from nitrates and phosphates was occurring in the part of the watershed where agriculture was taking place.

Concerned about Huntingdon's potable water supply, these sixth graders developed a series of recommendations for maintaining and improving the health of the stream. Then they got their message out by organizing public education events and making presentations to community members. All of their teachers—English language arts, science, history/social sciences, and math—guided the students along the way. In the end, these activities helped Huntingdon's students see both how statistical analysis works as a mathematical process and the role it can play in science and decision making.

Students are not the only ones who benefit from this approach; teachers find that using an I.I.I. strategy gives them greatly expanded opportunities to support and be supported by their colleagues. At Arabia Mountain High School in DeKalb County, Georgia, for example, the tenth-grade teachers function as an interdisciplinary grade-level team that works together to support instruction on the whole group of standards they have identified from the Common Core State Standards in English language arts and mathematics, as well as chemistry, history/social science, engineering, and healthcare sciences. For example, the chemistry teachers reinforce instruction in the mathematics standards, and the English language arts teachers focus on skills that the students will be using in history/social sciences, engineering, and healthcare sciences. When teachers work together on developing an EBE unit, they begin to share responsibility for student success. This encourages them to reinforce instruction in each other's content areas, and ensures that they are supporting the same writing strategies, teaching math skills students can use to analyze their own real-world scientific data, and directing students to use similar research strategies in all of their classes.

Arabia Mountain's teachers have also discovered that, as they collaborate with their colleagues, they learn new instructional strategies, such as hands-on activities, authentic and performance-based assessment, and service learning. What's more, they enjoy the time they spend collaborating because it provides them with intellectual stimulation as they learn about and become interested in the environmental issues within their own community.

DEVELOPING LEARNING OBJECTIVES

Once they have selected their environmental context and the standards on which they will focus, EBE planners are ready to advance into the process of developing learning objectives for their program. The decisions made during this process are critically important, because the learning objectives will drive the development of all program components, including the design of instructional activities, instructional materials, student assessments, and, ultimately, the summative assessments used to evaluate the effectiveness of the EBE program.

Environment-based education program planners define the term "learning objectives" in the same way as other educational programs: statements that describe, in specific and measurable terms, what a learner is expected to know or to be able to do as the result of participating in an instructional activity. Nonetheless, the focus of learning objectives in EBE programs differs from that of learning objectives written for traditional instruction.

The first goal of EBE learning objectives, like that of a traditional program, is to take students to proficiency with the standards. The difference is that EBE learning objectives are written from the perspective of the selected environmental context and are woven together through that context—the State Education and Environment Roundtable (SEER) calls these "context-based learning objectives." Since all EBE programs focus on natural and human social systems as well as standards, the systems-related content (e.g., components, processes, and interactions among systems) must also be incorporated into the learning objectives. In the ideal situation, the standards-based and systems-related content are woven together within many, if not all, of a program's learning objectives.

For example, after EEI consultants identified the target standards for California's EEI curriculum, they began the process of writing a series of learning objectives for each of the selected standards. They wrote these objectives to form an instructional progression that would provide students with the building blocks to take them to mastery of the standards and associated Environmental Principles (the context of California's EEI curriculum). The developmental progression of learning objectives for the tenth-grade EEI unit "Britain Solves a Problem and Creates the Industrial Revolution" interweaves instruction for California standards 10.3.1. and 10.3.5. with California's Environmental Principle I.

Development of Learning Objectives for EEI Unit

History/Social Science Standards

The goal of the tenth-grade EEI unit "Britain Solves a Problem and Creates the Industrial Revolution" included having students master the following history/social science standards:*

- 10.3.1. Students "analyze why England was the first country to industrialize."
- 10.3.5. Students "understand the connections among natural resources, entrepreneurship, labor, and capital in an industrial economy."

Environmental Principle and Concepts

Principle I, Concepts a and b (see appendix A at the back of the book)

Learning Objectives

Students will:

- recognize natural systems and the resources they provide (ecosystem goods and ecosystem services) as the basic capital for the development of an industrial economy.
- provide examples of the major connections between natural systems and resources, and entrepreneurship, labor, and capital in industrial economies (for example, the labor necessary to extract, harvest, transport, and produce ecosystem goods and ecosystem services for human communities).
- describe how increased demands provided an economic opportunity for the English to improve the methods they used to extract, harvest, transport, and produce goods from the natural resources that were available.
- recognize that the growth in human populations and human communities in England placed greater demands on natural systems.

* California Department of Education, *History–Social Science Content Standards for California Public Schools: Kindergarten Through Grade Twelve* (Sacramento: California Department of Education, 1998), 43.

The process of developing California's EEI curriculum demonstrates the importance of establishing learning objectives that are clear, specific, and agreed upon by the key stakeholders. The learning objectives drafted by the team were incorporated into the EEI Model Curriculum Plan, which was then reviewed and approved by the state agency partners represented on the Interagency Model Curriculum Planning Committee. Over the next six years, these approved learning objectives guided every aspect of the curriculum development process, including the phases of writing, editing, review, and ultimately approval by the State Board of Education.

Writers and editors, for example, were all reminded on numerous occasions that they had to view all of their work through the lens of the established learning objectives. Lesson plans, activities, readings, and assessments were all driven from the same perspective: "These learning objectives have been reviewed by the Interagency Model Curriculum Planning Committee, which has determined that if students can demonstrate the specified knowledge and skills as a result of the EEI curriculum, they will be deemed as having achieved 'mastery' of the standards." In this way, the learning objectives served as "guideposts" throughout the developmental phase against which the work of the writers and editors could be measured. Ultimately, of course, these learning objectives also formed the basis for assessing students' academic success.

Even though the scale is substantially different, the process of developing learning objectives for EBE programs in schools and school districts is the same as it was for developing California's EEI curriculum. The fifth-grade "Water Works" unit at Jackson Elementary School in Altadena, California, exemplifies learning objectives developed for teaching students standards in English language arts, math, science, and history/social science through a community-based investigation of water consumption at the school and in the local area. The teachers wanted to create learning objectives that would, for example, teach two science standards: "Students know that the amount of fresh water located in rivers, lakes, underground sources, and glaciers is limited and that its availability can be extended by recycling and decreasing the use of water" (5.3.d.); and "Students know the origin of the water used by their local communities" (5.3.e.).[9]

Since Jackson's teachers also wanted their students to learn about local natural and human social systems, they incorporated this content into a series of context-based learning objectives. For example: "Students will describe how water is supplied to their school and how it is utilized at school"; "Students will describe the flow of water into the school and the flow of water waste out of the school and the community at large"; "Students will identify interactions among the school community and the natural systems in and around the school"; and "Students will make predictions about the effects of natural and social interactions in and around the school community." These context-based learning objectives gave teachers a platform for teaching both the targeted standards and the systems-related content. (For additional learning objectives in this unit and for a cross-section of the learning objectives in the Seven Generations Charter School's second-grade "Fantastic Forests" unit, see the appendixes at the end of chapter 5, p. 115)

Having completed the process of selecting their standards, establishing interdisciplinary connections, and writing learning objectives, EBE planners have their instructional framework in place and are ready to begin developing and field-testing their lesson plans and instructional materials.

Thinking About Learning Objectives

When writing learning objectives, keep in mind that collectively they should:

- lead students to proficiency with the standards;
- be written from the perspective of the selected environmental context and woven together through that context;
- cover content related to environmental principles, concepts, and topics so students will be examining the interactions between natural and human social systems;
- interweave and interconnect the environmental context and content with the standards-based content;
- help students see connections to their daily lives, future jobs, college choices, careers, and their communities;
- incorporate environmental content into measures of student achievement.

CHAPTER EIGHT

EBE Instructional Materials and Resources

Natural systems and human social systems and their interactions provide a diverse tapestry of learning opportunities. The breadth of content encompassed by environment-based education (EBE) programs often demands the involvement of individuals and teams with wide-ranging expertise in the process of developing the instructional programs and related materials. Fortunately, most communities have historians, scientists, government agencies and officials, museums, nature centers, community groups, businesses, and individuals available to lend their expertise and educational resources to support environment-based education. Our experience and observations over the past thirty years have shown that these individuals, agencies, and organizations are very interested in this topic area and are willing to invest their time and resources in supporting EBE programs and making them a success.

The development of programs at individual schools or districts can often be completed by teachers and other curriculum writers with the support of individuals who have technical expertise related to environmental content and the local community. Developing large programs like California's Education and the Environment Initiative (EEI), on the other hand, can require hundreds of consultants, writers, editors, cartographers, photographers, and graphic designers, as well as numerous teachers and their students for field-testing of draft materials.

Environment-based education program planners are ready to begin the process of writing lesson plans, developing supporting materials, and identifying instructional resources once they have all of the key components of their instructional framework in place: program goals; environmental context; academic standards and systems-related content; decisions about interdisciplinary connections; and learning objectives focused on the selected standards and systems content.

The specifics of the framework drive the process of developing the lessons and supporting materials. A program that is focused on a local environmental context, for example, generally requires participating educators to develop or locate lessons that are directly relevant to their local environment, while instructional materials designed for a context based on big environmental ideas are more likely to be relevant across larger geographical areas. As an example of the latter, California's EEI curriculum is applicable across the state, and many of its components could easily be adapted to other states as well.

This chapter identifies each of the major instructional components of an EBE program, describes the most important characteristics of each component, and discusses how these resources can be developed or located. It then summarizes the major developmental stages of creating, reviewing, testing, producing, gaining approval, and disseminating the program materials.

KEY INSTRUCTIONAL COMPONENTS

Environment-based education programs are organized into instructional units that are framed around a series of sequential learning objectives and designed to teach one or more academic standards in combination with content about natural and human social systems. Whether focused on the context of a local environment or a big environmental idea, each instructional unit comprises three major components: lesson plans, supporting materials, and implementation resources. There are, however, several differences between each of the components within these two types of contexts.

Lesson Plans

The lesson plans that are used in an EBE instructional program are structurally the same as those used in a more traditional approach to classroom teaching. In both cases, the key components of a good lesson plan are learning objectives that describe: measurable tasks; teaching and learning activities; and strategies to assess student progress toward achieving the learning objectives.

Environment-based education instructional units are typically composed of three different types of lessons: subject-specific, systems-related, and blended. Figure 8.1 provides a graphic representation of the instructional framework and expected outcomes of each of the three types of lessons in an EBE instructional unit.

When there is a need for *subject-specific lessons*, they should cover content that is directly focused on the development of basic skills and content knowledge—for

FIGURE 8 .1
Lesson flow in an EBE instructional unit

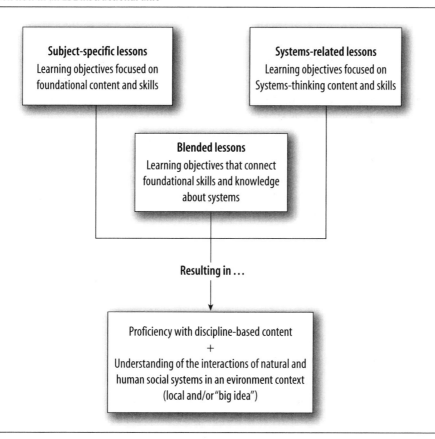

example, foundational reading and writing skills, math operations, aspects of earth sciences like weathering and erosion, and history/social science skills such as making and interpreting maps. *Systems-related lessons* are tied to the development of core understandings of the components, processes, cycles, and interactions within systems, such as direct and indirect changes to natural systems resulting from consumption of natural resources by humans. The third group, *blended lessons,* are those designed to focus on the learning objectives that bring together the standards-based and systems-related content. They can also be used to make interdisciplinary connections, such as by combining history/social science content with that of science and English language arts.

Most of the lessons used in EBE instructional units that are focused on a local environmental context are created by participating teachers. These teachers often supplement their own lessons with others that have been developed by entities familiar with their local environment and community, published by state and national environmental organizations, or that are from their adopted instructional materials. (Over the past ten years, we have seen several teachers become so invested in the process of developing their EBE instructional units that they have refined them and used them as part of their master's and doctoral research programs.)

Educators should carefully review any lessons from other organizations before incorporating them into their own EBE instructional units. Lessons from all sources should be carefully assessed for bias, relevance to students, technical accuracy of the information, and, unquestionably, their focus on the standards targeted for the unit and the effectiveness of the included teaching strategies. Sometimes, for example, environmental organizations produce lessons that reflect a particular point of view, and in many cases, students view activities from adopted instructional materials as irrelevant to their daily lives and their communities. And unfortunately, lessons that are readily available may or may not be age-appropriate, and to make them useful may require extensive adaptation of reading level or topical focus.

Environment-based education program developers often resolve these issues by extensively adapting or rewriting some of the existing lesson materials that they discover. In their Environment as an Integrating Context for learning (EIC) Model "Watersheds" unit, for example, the middle school teachers at Seven Generations Charter School in Emmaus, Pennsylvania, used a combination of several self-developed lessons and others that they adapted from various sources, including

WET in the City and lessons from California's EEI curriculum.[1] There are diverse national, state, and local organizations and agencies that make their materials available on the Internet. For example, teachers at Seven Generations discovered rich resources that had been developed by the Delaware Riverkeeper Network, Philadelphia Water Department, and even the Oklahoma Biological Survey. (Note that it is extremely important to get any needed permissions prior to reproducing any lessons or other materials that are protected by copyright. Many organizations make their materials available at no cost. For example, all of California's EEI curriculum units are available for downloading at no cost.[2])

In many cases, because of these concerns and others, it may be necessary to develop completely new and original lessons and materials for a given program. For example, before undertaking the development of the California EEI curriculum, the state's team of consultants conducted a nationwide search for relevant, existing lesson plans. The team collected and analyzed over 200 samples of instructional materials from local, state, and national entities for their potential use in conjunction with this curriculum. None of these lesson plans was found to be appropriate. Some were not focused on standards-based learning, and a few reflected a biased point of view and/or had no connection to the state's Environmental Principles and Concepts. (Several lessons developed by California's San Joaquin County Office of Education, although standards-focused and unbiased, had not yet been connected to the principles.)

The tenth-grade unit "Britain Solves a Problem and Creates the Industrial Revolution," like all the other units developed for California's EEI curriculum, is a completely original collection of standards-driven lessons, student activities, and assessments.[3] The development of this unit was necessary because none of the history/social science textbooks adopted by districts across the state contained any lessons that covered Britain's Industrial Revolution from the perspective of the interplay among natural systems and resources, entrepreneurship, labor, and capital in the developing industrial economy. Further, none of the existing lessons then in use in California's tenth-grade programs connected instruction to the Environmental Principles and Concepts.

There are a few other items to remember when developing EBE instructional units. Community mapping activities are only one of the many ways to get students engaged in instruction that is outside of their classrooms. Environment-based education programs using the local environment as context should build in as many activities as possible that give students hands-on opportunities to practice skills and expand their content knowledge as they learn about their local environment and community surroundings. Similarly, programs that focus on big environmental ideas can be made much more effective if they, too, are connected with real-world learning experiences. California's EEI, though to a great extent a textbooklike instructional approach, includes numerous hands-on activities and many recommended "extension ideas" to give students an opportunity to take what they are learning in class out into their communities. Environmental service-learning projects can provide many different opportunities for students to learn and further apply what they are studying and at the same time make a meaningful contribution to their community. Activities like these reinforce students' knowledge, help them further develop many of their twenty-first-century skills, and engender a sense of civic responsibility.

Supporting Materials

Environment-based education instructional units frequently use everything from written to visual and even audio materials to enhance students' learning and interest in the environmental context. Program developers quickly discover that they can find, adapt, or create an almost unlimited array of resources to support their lessons, whether they are involved in studying local environments or big environmental ideas. They find that they can easily build good working relationships with organizations like historical societies, nature centers, museums, research institutions, businesses, agencies, and many others. Almost without exception, these groups are willing to share everything from student readings to scientific research publications, maps, and many other visual aids like charts, videos, and PowerPoint presentations. Many of these agencies and organizations even have this type of public outreach as an important part of their operating responsibilities.

Educators who are developing an EBE program around a local environmental context for the first time are often surprised by the extensive range of resources available from local, state, and national agencies, as well as universities, environmental research facilities, local organizations, nature centers, the Internet, and both public and university libraries. Water quality data, for example, are plentiful and easy to find, because they are collected by local, regional, state, and federal natural resource and environmental agencies. With just a few minutes of research, students can readily find data like these, most of which are now posted on the Internet. Similar information about other topics—like air quality, energy consumption and production, fish and wildlife populations, local and state parks, wildlife refuges, state and national forests, and many others—can be readily located either on the Internet or by making a few calls to the offices of the appropriate agencies or organizations. Students and teachers soon discover the types of information they need, who collects it, and how they can find it. This is equally true for maps, aerial photographs, satellite images, and many other types of visual materials that bring EBE programs to life.

As they were developing and implementing their "Water Works" unit, Jackson Elementary School teachers Peters and Silverio gathered and adapted supporting materials from many local and state organizations. The teachers pointed students to the Internet when they began looking for possible solutions to water waste on the campus, at their homes, and in the local community. They discovered that the Metropolitan Water District's Web site contained especially useful information, as did other educational resources available from the Los Angeles County Department of Public Works, the Altadena Town Council, and the Altadena Historical Society.

Some EBE programs find that they must create all of their own supporting materials, either because they are not otherwise available or the organizations who own the copyright will not give permission for their use. The supporting materials for "Britain Solves a Problem and Creates the Industrial Revolution" were extensive, and required the writing of several stories, the development of vocabulary lists, and the creation of a series of "information" cards, student activity sheets, and even a "political" map, which was part of a series developed specifically for the EEI curriculum by the National Geographic Society. Other EEI curriculum units include everything from games and puzzles to audio and video clips, posters, and even specially developed books, like *California's Natural Regions*, that contain material not otherwise available for third- and fourth-grade students.[4]

Most EBE programs follow the path taken by the teachers at Jackson Elementary School, an approach that provided many rich learning experiences at little or no cost. California's EEI curriculum is an exception to this general rule, because it was well funded and was developed for statewide distribution.

Implementation Resources and Support

The public's awareness of, interest in, and concerns about the environment have grown dramatically over the past several decades, starting with the 1962 publication of Rachel Carson's *Silent Spring*, and have expanded more recently following the publication of former vice president Al Gore's book *An Inconvenient Truth*.[5] The growing consciousness of local and global environmental issues has resulted in increased commitments from individuals and organizations willing to support and participate in efforts that help young people become more aware of these issues. In addition to the background information they may provide for teachers as supporting materials, these community partners can also help with a variety of instructional activities both on campus and away from the school.

Our experience with EBE programs across the United States provides extensive evidence and examples of individuals, businesses, nongovernmental organizations, universities, and many others that have become actively involved in these programs. Appendix D, at the back of the book, lists many of the individuals and agencies that have supported the EBE programs discussed throughout this book. Seven Generations Charter School's first-grade teaching team of Jennifer Hersh, Angela Waldraff, and Mara Richardson, for example, in the process of implementing their three instructional units, worked with thirteen of the fifty community partners that support the school's work.

This support is crucial to the successful implementation of EBE programs, whether at the local level or on a larger scale. Although the types and extent of support that are needed vary substantially among programs, there are three general kinds: logistical assistance, technical expertise, and program support.

The variety of logistical assistance that EBE programs need at the local level ranges from simple things like parents helping to monitor hands-on classroom activities and the planting of campus gardens, to more extensive involvement like helping students build greenhouses and helping teachers supervise groups of students when they are off campus participating in their community-based investigations. Parents, senior volunteers, employees of local businesses, and even school and district staff members like facilities managers often have an integral part in campus and community activities.

The educators at Seven Generations Charter School in Emmaus, Pennsylvania, regularly engage community members as part of their logistical support team. In a third-grade "Interdependence on South Mountain" unit, for example, parents at Seven Generations Charter School worked alongside students, teachers, volunteers from local businesses, and a native-plant specialist to implement an environmental service-learning project to restore the riparian zone along the creek that runs from the mountain and down past the campus. The native-plant specialist visited the students and told them about the differences between native species and invasive plants. The parents joined in the process by helping students collect the litter in the creek and remove the invasive plants. Employees from a local nursery met with students and discussed how to select the most appropriate species for their restoration project. Finally, as a large team, students, teachers, school staff members, parents, and other community volunteers began the restoration process by making repairs where soil had eroded from the banks and replacing the invasive species with native plants. While a school or district's educators are responsible for most of the work involved in developing and implementing EBE programs in a local context, they benefit greatly from bringing in members of the community to support their efforts.

Not surprisingly, large-scale EBE programs, like the multischool program at Desert Sands Unified School District in Southern California's Coachella Valley, need even more assistance with logistics. Coordinating all of the organizations and individuals involved in a big program is a major responsibility, so these programs frequently assign one or more individuals to this task. Desert Sands had several individuals involved in program coordination: one responsible for working on the educational aspects of implementation; another who was the district's resource conservation specialist (waste reduction and recycling were an important aspect of their education program), and a teacher-coordinator assigned for each of the six participating schools.

There are several other forms of help that, depending on the specifics of the program, may be required to successfully implement an EBE program. Among them are: permission from those who manage parks, rivers, lakes, forests, farms, and businesses to visit and use facilities for off-campus field experiences and community-based investigations; access to technical experts who can review scientific or historical aspects of students' research and give them guidance; introductions to decision makers who might be able to provide financial or "political" support; access to analytical equipment for collecting and analyzing data like water and soil samples (e.g., as was supplied by Juniata College for Huntingdon Area Middle School students); and many others. Some state and local agencies maintain directories of environmental organizations and resources that can benefit educators implementing EBE programs. One example is the California Regional Environmental Education Community (CREEC) Network, which was developed and is maintained by the California Department of Education.[6]

There is one caveat about getting assistance and support from all of these different individuals and entities: they must understand that you are asking them to support your efforts to achieve the vision, mission, and goals that your team has established. They are likely to have their own set of goals, programs, and materials, which may or may not meet your needs. This often means that you will be asking partnering individuals and groups to adapt or develop new materials for your students, rather than implementing what they might otherwise have been using for many years. (This is especially important if you are working with individuals who are not experienced with standards-based education or are not up-to-date on your area's most current standards.) These issues can be minimized if your potential partners are involved in program planning so that they clearly understand your intentions and the focus of your efforts.

DEVELOPMENT PROCESSES

This section focuses on the major processes involved in creating high-quality EBE programs and materials—from draft versions through dissemination of final materials. These same steps should be applied to the development of all EBE programs, whether they involve a few classrooms, a school, a school district, or a whole state. Even if they must be applied to a lesser degree due to limited resources, the same processes and considerations apply to smaller-scale programs, since whatever the size or scope of the program, the goal is to successfully create, gain approval for, and disseminate high-quality EBE program materials.

Given that the context, standards, and learning objectives for the program have been established, there are seven major phases involved in developing EBE instructional programs and materials, including:

1. Assessing needs and planned use
2. Writing and editing
3. Technical review
4. Field-testing and revision
5. Graphic design and production
6. Administrative review and approval
7. Use and dissemination

The reaction of many educators to this list of developmental processes may be, "I have never before had to go through all of this with any programs that I have created." This is a reaction to the difference between programs developed for use in a single classroom and programs that are intended to achieve wider distribution and use in multiple classrooms, or in a whole school, district, or state. Ultimately, these stepped processes increase the quality of a program, its effectiveness, and its credibility with the administrators and policy makers who will, at the end of the day, control the implementation of an EBE program.

Key Process Steps in Developing Instructional Materials and Programs for California's EEI

Phase 1—Assessing Needs and Planned Use

- Determining what educators need and how they will use the program
- Designing materials templates based on the needs assessment

Phase 2—Writing and Editing

- Selecting skilled writers and editors to develop unit descriptions based on the design template and established learning objectives
- Writing drafts based on the design template and learning objectives
- Reviewing drafts for content, mechanics, completeness, and conformity with established learning objectives
- Revising and preparing drafts for technical review

(continues)

Phase 3—Technical Review

- Identifying highly qualified subject matter experts
- Having technical experts review drafts
- Revising materials based on experts' reviews
- Preparing draft materials for field-testing

Phase 4—Field-testing and Revision to Enhance Quality

- Distributing materials for field-testing by teachers
- Gathering and analyzing results of field-testing
- Revising materials based on field-test results

Phase 5—Graphic Design and Production

- Designing instructional materials, including text, diagrams, illustrations, and maps
- Producing materials for review and approval

Phase 6—Getting It Approved

- Learning about the formal and informal approval processes that will be used to review the program
- Obtaining approval for implementation from appropriate administrators and policy makers

Phase 7—Dissemination and Professional Development

- Developing strategies for getting the program disseminated and widely used
- Building awareness of the program among teachers, administrators, and policy makers by achieving and publicizing early successes

The discussion that follows describes these phases and reviews some of the most significant issues that are encountered in developing programs ranging from the size of California's EEI to smaller programs like those at Jackson Elementary School and Seven Generations Charter School. While few agencies, organizations, or individuals will be working to develop a program as large as California's EEI, there is, nonetheless, a great deal to be learned from that process.

ASSESSING THE NEEDS OF EDUCATORS

Planners of successful EBE programs always approach their task from the perspective of their "end users." Whether discussing their work with stakeholders, writers, editors, technical reviewers, graphic designers, or decision makers, they always focus on the two most critical questions: "Will the program and instructional materials we are creating benefit students?" and "Can what we are developing be readily implemented by teachers?" In addition to helping improve the program's quality and effectiveness, asking these questions of teachers and education administrators at every stage of the design, development, and review process gives them an opportunity to be actively involved, and can eventually engage them as willing "ambassadors" during the dissemination phase of the program.

Every teacher, school, district, and state has its own approaches to instructional design, and will differ in the ways that they intend to use an EBE program and its associated lessons and supporting materials. These differences arise as a result of the way teachers are trained, how administrators of districts formulate their programs, and how state education leaders structure their strategies for implementing standards and their statewide assessments. The use of instructional materials varies as well, but there are twenty or more states that adopt materials at a state level, although the choices among textbooks and supporting materials are driven principally by four states: California, Florida, New York, and Texas.

In view of these factors, it is important to gather input from teachers and administrators regarding the specific designs that should be used to formulate instructional materials and programs. These data can be collected in many ways, including through interviews, written surveys, focus-group sessions, and committee meetings. Written surveys are one of the most effective means of gathering information from large stakeholder groups like teachers.

California's EEI implementation team undertook a written "Educator Needs Assessment" prior to designing the EEI curriculum.[7] The survey was distributed to 9,657 teachers who were randomly selected from the active rolls of the California State Teachers' Retirement System. Information from the 361 teachers who replied proved valuable throughout the design and development process.

The teachers at Jackson Elementary School and Seven Generations Charter School used SEER's EIC Model planning template as a framework for developing their instructional units. They then used the framework that they created in the unit template to write their lesson plans and to gather supporting materials and any implementation resources they might need.

Examples of Questions Used for the EEI Educator Needs Assessment

- How do you use state adopted instructional materials?
- How do you use supplemental instructional materials?
- How much time do you allocate each day for using adopted and supplemental materials?
- In what format(s) would you prefer that curriculum materials be developed by the EEI?
- What types of instructional materials do you want?
- What criteria are most important in choosing which instructional materials you will use?
- What grade levels do you recommend as the focus for developing the EEI curriculum?
- How likely are you to use the EEI curriculum to teach standards in different disciplines?

Key Elements of SEER's EIC Model Planning Template

- Community map
- Description of environmental context, including natural and human social systems and their interactions
- Academic content standards and learning objectives
- Systems-related content and learning objectives
- Description of community-based investigation(s), including organizing and supporting questions
- Discussion of lesson plans (subject-specific, systems-related, blended, and community-based investigations), including: descriptions, preassessments, prerequisite knowledge, student assessments, supporting materials, and implementation resources
- Description of environmental service-learning activities
- Implementation timeline

Writing and Editing Materials

The time needed to write instructional materials for EBE programs varies greatly, depending on the scale of the endeavor—an EIC Model unit can be developed in a matter of weeks, but developing the EEI curriculum required four years. At whatever scale, these efforts can be facilitated by carefully designing the structure of the materials before writing begins, by involving individuals who have experience writing standards-based curriculum, and by seeking ongoing guidance from editors, program managers, or writing mentors.

The first step requires the development of a uniform outline and plan to guide all participating writers. The format and structure should be designed in response to the information gained during the assessment of educators needs, and augmented by any input received from school, district, or state administrators. The writing of templates should encompass and describe all critical parameters for an instructional unit, including individual lessons, background material for teachers and students, assessments, readings, visual aids, and also details such as the reading level at which all the student materials should be written. Large-scale programs that are going to produce formally published materials usually require very detailed templates that even specify items such as word counts allowed in each section.

Identifying and engaging experienced writers can be a difficult process, but it is crucial to developing a standards-based program that conforms to specific design parameters. It is important that the program writers—whether individual teachers in a school or professional writers—understand and are agreeable to conforming to the program's strategy for materials development. It will be helpful when selecting writers and editors to ask for writing samples specifically created to match your program's unit templates, since it is not always easy to determine from a writer's previous work whether he or she has the skills necessary for this particular task.

Editors, whether focused on instructional quality or grammar, can play a very important role in ensuring that materials are well written and complete. The California EEI used several different types of editors, including individuals responsible for checking instructional strategies, technical content, and grammar. The teaching teams at Jackson Elementary School and Seven Generations Charter School had SEER's staff members review the draft materials for their instructional units and make recommendations for improvements.

Technical Review

External technical experts should thoroughly read and check the accuracy of the scientific, historical, and technical aspects of all draft instructional materials. Academic experts and members of state and local environmental and historical groups and agencies are all good resources for this task. It is a good idea to begin the search for well-qualified technical reviewers early in the development process, and then set aside adequate time for them to complete thorough reviews.

Small-scale programs such as those at Jackson Elementary School and Seven Generations Charter School typically do not have funding for technical reviews. In these cases, the best option is to find unpaid reviewers from local environmental agencies, universities, or nonprofit organizations, or colleagues and parents with specific training and knowledge. This type of technical review and advice is an important contribution that community partners with specific technical expertise can make to their local schools.

In California's case, the curriculum development team originally intended to have two external experts review each of the units. In spite of the stipend offered, however, not enough qualified applicants applied (perhaps understandably, since individuals with knowledge about ancient Egyptian and Mesopotamian farming practices—a history/social science standard for California sixth graders—may not be a common commodity). Although it took substantial effort to find them, each EEI curriculum unit was reviewed by at least three knowledgeable experts.

Field Testing

Field testing of instructional materials is a critical piece of the development process, offering the opportunity to assess the practical aspects of all components of the program's instructional materials, including: adequacy of background information, clarity of instructional procedures, utility of visual aids and reading materials, suitability for intended grade level, and efficacy of content, procedures, and assessments for targeted standards. Equally important, field testing gives instructional materials increased credibility with other teachers and educational administrators.

The teachers involved in small-scale EBE programs also need to undertake field testing, by gathering results and planning revisions based on their experiences using each unit in their classrooms. At schools like Jackson Elementary and Seven Generations, after they have implemented the unit in their classrooms, the teachers bring their results to other members of their teaching team. They share their experiences and work together to revise their plans so that they can use the materials the following year.

California field-tested the EEI curriculum in schools that were representative of the state's diverse demographics. The process began with sorting through all 1,057 school districts in the state to identify those that might be willing to participate. Eventually the field testing was focused on twenty districts, involving over 200 teachers and almost 5,000 K–12 students. The teachers fit the EEI units into their already-planned annual instructional schedules. This approach gathered information about the effectiveness of the curriculum and teachers' suggestions for revisions. Of the five teachers who tested each unit, one took on the responsibility of consolidating the feedback and submitting the group's recommendations to the curriculum development team so that appropriate revisions could be made to the draft documents.

Graphic Design and Production

The curriculum materials created for EBE programs must compete for classroom time and student interest with the engaging graphics produced by large commercial publishers. If EBE programs are to succeed, their instructional design must meet or exceed these same standards.

The process of preparing EBE instructional materials for graphic production involves several steps, including creating unified templates for graphic design, identifying potential sources for these graphical elements, and establishing a multilayered review process. The steps in this process are the same across all programs, regardless of size. Some of the most important curriculum design components are: teachers' guides, teachers' masters, visual aids, student workbooks, student readers, student dictionaries, maps, and other supporting materials like word-wall cards for the early grades.

Getting Approval

The particular procedure for seeking and getting approval to implement an EBE program and its associated instructional materials depends on the workings of any given school, school district, or state education system. This makes it very much worth the effort to become fully informed on whatever review and approval processes will be applied to a newly developing program. All too often, individuals and organizations make the major investments involved in creating instructional programs that are ultimately rejected by key decision makers because the developers did not follow appropriate procedures.

Large-scale programs are typically subject to very formal review and approval processes that usually require many months and involve numerous reviews by diverse groups, including the public, appointed review committees, and the school district or state education agency in charge. For smaller-scale programs, such as at the individual school or classroom level, principals may have blanket authority to sign off on new instructional programming. (That individual principals have the authority to make such decisions has become much less common, though, as school districts have responded to the growing demands of statewide student assessments.)

The process of getting an EBE program reviewed and approved at the state level can be quite cumbersome, but achieving this level of recognition is unquestionably worth the effort. This investment can pay multiple dividends, including winning credibility with district and school administrators, as well as with teachers, parents, and funders.

Access to additional sources of potential funding is another advantage of gaining approval through a formal review process. Most funders, whether governmental, corporate, or philanthropic, will not award major grants unless they can verify a program has already been approved for use in schools. And recent cutbacks in educational budgets at all levels make such alternative funding sources even more important.

The East Penn School District, the entity responsible for chartering the Seven Generations Charter School in Emmaus, Pennsylvania, has authority for reviewing Seven Generations' programs and curriculum. School district administrators reviewed a draft version of the school's instructional plans at the time they granted the school's original charter application and again three years later when determining if they would renew the charter.

At Jackson Elementary School in Altadena, California, the teachers did not need formal approval of their instructional program or materials from their school district. The school's principal did, however, review and approve the classroom use of the fifth-grade "Water Works" unit and the six-grade "Reducing Paper Usage" unit.

Planning for the process of formal review and approval of California's EEI began very early in development, because approval of the program curricula needed to follow the same procedures used for traditional textbooks. Because of all the procedural requirements, the schedule for this work was established about two years before it actually began. Ultimately, though, this investment of effort resulted in the State Board of Education's formal approval of the EEI curriculum. This approval enables district educators to use these instructional materials as replacements for state-adopted textbooks—substantially increasing the confidence of end users like district administrators and teachers, who are mandated to ensure that 90 percent of materials used in the classroom are state approved.

Dissemination

It is a major accomplishment to complete the plans for implementing an EBE program through all the phases that lead to the production and approval of instructional materials, but this is not the end of the process. Similar effort must be dedicated to developing strategies for getting the program widely used and disseminated.

Regardless of the scale of the program, the ultimate goal of all the work on development and approval is to get EBE programming and materials into the hands of teachers and students. It is important to keep this goal in mind throughout the planning process and to maintain an appropriate balance of investment among the stages of designing, developing, producing, and disseminating the program.

Large-scale endeavors, like the EEI, often invest the majority of their resources in the development of the instructional program materials, seeking to avoid what might become prohibitively high costs for printing, dissemination, and professional development. At the other end of the spectrum, smaller efforts that are focused within individual schools and school districts require greater expenditures on professional development, but relatively little for materials development and production.

The principal investments at Jackson Elementary School and Seven Generations Charter School, for example, involved the participation of teachers in one of SEER's EIC Model professional development institutes, followed by instructional planning time. These teachers drafted their own local environment–specific instructional units with a great deal of individual effort and team work. This level of effort gave them a very strong sense of ownership, which fueled their commitment to implementing EBE programs in their individual classrooms. It also made them enthusiastic about sharing what they had developed with their colleagues in other classrooms and at other grade levels within their schools.

In some cases, dissemination becomes a more formal part of the process. Administrators in the East Penn School District, for example, built dissemination work into the activities they are requiring as part of the charter renewal for Seven Generations Charter School, and over the next several years, Seven Generations will be conducting discussion groups about EBE with other local school districts.

Similarly, many schools in SEER's EIC Model School Networks have hosted visitors and observers from other schools and districts interested in learning more about EBE.

California's EEI program had to take a different approach, because the law that created it only directed that the curriculum was to be posted and maintained on the state Department of Education's Internet site, and made no provisions for printing or professional development. Because of the vast scope of the EEI curriculum, the California Environmental Protection Agency decided that this would not be sufficient for getting the materials disseminated and into widespread use. The EEI implementation team examined several alternatives to traditional professional development and dissemination of printed materials. They chose to focus on an online strategy, and produced a series of Web-based course modules that introduce teachers to all of the important parts of the curriculum. Additionally, in order to avoid what could become unsupportable printing costs, the state made arrangements with a commercial printing firm that was willing to publish the materials using a "print-on-demand" strategy—a very important option in light of widespread issues related to education budgets.

Finally, in developing a dissemination strategy, it is important to focus on achieving and publicizing early successes. These efforts should lead to growing awareness of the program by teachers, administrators, and policy makers, and should eventually result in the EBE program being used in more schools and classrooms.

The final chapter discusses the two last key elements of every successful EBE program—student assessment and program evaluation. It also presents examples of how these tools can be used in the ongoing efforts to improve and expand EBE programs.

CHAPTER NINE

Student Assessment and Program Evaluation

How do we know environment-based education (EBE) programs work? The research cited in chapter 3 strongly indicates that students' achievement is enhanced in EBE programs. The research base is constantly growing, because many EBE planners and teachers are evaluating their work, and that of their students, along the way.

Every professional educator is well aware that a crucial part of all good lessons and instructional units is a student assessment component that gauges progress toward the established learning objectives. Environment-based education programs, like all other high-quality educational approaches, build in student assessment as a part of each instructional plan. As such, EBE instructional materials include: components for preassessment; remediation plans as needed so that students have all necessary prerequisite knowledge; and postlesson and/or postunit student assessment instruments. In some instances, these tools may be relatively informal, and at other times they will be much more structured. For example, Seven Generations Charter School in Emmaus, Pennsylvania, in developing units based on the Environment as an Integrating Context for learning (EIC) Model, includes all of these features, as well as an assessment of the community-based investigations that encompass all of a unit's learning objectives, in each of its standards-based and systems-related lesson plans.

Evaluation of implementation activities is also important, and serves a function similar to the role that student assessment plays during teaching. Just as a lesson plan is not complete without assessment, an EBE program implementation plan is not complete without an evaluation component. The evaluation strategy and type of data to be collected derive directly from each program's decisions about its vision, mission, and goals. The data that are collected then serve as the basis for analyzing what needs to be done to continuously improve and strengthen the EBE program. They help to answer program-level questions like: "Are teachers adequately prepared to design and implement the instruction?" "What additional technical expertise and resources would strengthen the program?" and even "Are student assessments being built into every lesson and unit?"

Since both student assessment data and program evaluation data are useful tools for improving instruction and strengthening implementation, they should be gathered throughout the process of developing and implementing EBE programs, and not just serve a summative function as evidence to include in a "final" program report or study. Of course, as important as it is to collect and use these data as formative tools, it is important to balance the time and effort spent on data gathering and analysis with other aspects of program implementation. As a rule of thumb, it should be noted that many funding agencies have added a requirement for assessment and evaluation into their grant applications, setting 10 percent of overall program costs as the appropriate level.

This final chapter discusses and presents examples of plans and instruments used by EBE programs for assessing student progress and conducting formative program evaluations. The student assessment section looks at ways that EBE programs determine how they affect standards-based learning, higher-level thinking skills, and growth in students' knowledge about the environment. Program evaluation strategies are presented and discussed as they apply to monitoring the development of an EBE program in terms of its instructional materials, instructional methods, and the availability of implementation resources.

COMBINING TRADITIONAL AND ALTERNATIVE STUDENT ASSESSMENT

Student assessment is an integral part of all EBE programs, as it should be with all standards-driven education. The type of learning objectives that a team is emphasizing—subject-specific, systems-related, or blended—will dictate the appropriate type of assessment to use—traditional assessment, authentic/performance-based strategies, or a combination of both. Whichever of these approaches an EBE program team chooses to use, what is important is that both standards-based learning and systems-related learning are assessed, and that the resulting information is used to improve instruction.

A unique benefit of EBE is that instruction designed around environmental contexts offers diverse opportunities for students to demonstrate their knowledge in many subject areas through their work on community-based investigations and environmental service-learning projects. The students' real-world activities play such an important role at Georgia's Arabia Mountain High School, for example, that the teachers use them to assess proficiency with the standards as part of an overall assessment strategy. This type of combination strategy provides teachers with three different approaches to determining student progress: traditional tests, state assessments, and evaluating students' service-learning projects. As Arabia Mountain's Assistant Principal Tim Wells described it, "Students are expected to articulate and discuss which standards they applied during their projects, and how they used this knowledge and skills during their investigations. The students are able to see the standards come alive within their projects, which helps them achieve a higher level of success with the standards."

Environment-based education programs also measure students' knowledge of the environment as it relates to the components, processes, and cycles within natural and human social systems. In some cases, EBE educators assess this environmental knowledge at the same time as they assess the growth of students' twenty-first-century skills. Students' local investigations and projects are an excellent venue for authentic assessment of their critical-thinking, problem-solving, self-direction, collaboration, adaptability, leadership, and communication skills.

The teachers at Seven Generations Charter School are very interested in monitoring their students' development of these important skills, especially those of critical thinking and problem solving. Because they place such great emphasis on their local natural and human social systems and the effects on the environment that result from the interactions among these systems, they have developed two assessment instruments related to these skills. A team of Seven Generations teachers developed a critical-thinking and problem-solving rubric that they use toward the end of each instructional unit to assess the development of students' critical-thinking skills and their understanding about systems and environmental concerns related to systems (see appendix: "Seven Generations Charter School Critical-Thinking and Problem-Solving Rubric," at the end of this chapter). The problem-solving portion of this rubric helps teachers assess whether students understand the environmental topics they are investigating and can identify and explain possible solutions for these issues.

With its more textbooklike structure, each unit in California's Education and the Environment Initiative (EEI) curriculum includes instruments and instructions for both traditional and alternative assessments, giving teachers the option to choose whichever they consider most appropriate. The traditional instruments are designed to resemble the state's standardized assessments and, as such, depending on grade level, are composed of multiple-choice, fill-in-the-blank, short-answer, and sequencing questions. The alternative assessments, on the other hand, give students a choice of performance-based approaches for demonstrating their learning, such as completing projects that allow them to show their capacity for applying their knowledge and skills to real-world problems in their local communities. Examples of these assessment tasks include creating and recording public service announcements, designing posters, writing editorials for newspapers, making and writing postcards, and creating concept maps that describe all the interlocking parts of environmental problems as they apply to natural and human social systems.

Even though it took a more traditional approach to instruction and assessment, the EEI curriculum built in instruments that could determine how students were progressing in terms of their knowledge of the environment in the context of interactions between natural and human social systems. In this case, the assessment of students' systems-related knowledge was fully integrated into the assessment of the history/social science and science standards that were the focus of each EEI unit. For example, both the end-of-unit and lesson-level assessments for "Britain Solves a Problem and Creates the Industrial Revolution" demonstrate how a traditional instrument can combine the assessment of standards- and systems-related knowledge.[1]

In "Britain Solves a Problem," these kinds of learning objectives and lesson-level activities lead to a written unit assessment that may at first appear traditional, but in terms of the content that it covers, it is very different from a typical tenth-grade world history assessment. The unit-level assessment, for example, asks students to do things like: "Explain why a country's natural systems and the resources

Britain Solves a Problem and Creates the Industrial Revolution

Tenth-Grade EEI World History Unit

Summary of Learning Objectives Driving the Assessment of Standards and Systems

Standards

10.3.1. "Analyze why England was the first country to industrialize."

10.3.5. "Understand the connections among natural resources, entrepreneurship, labor, and capital in an industrial economy."

LEARNING OBJECTIVES	SUMMARY OF UNIT- AND LESSON-LEVEL ASSESSMENTS
Recognize natural systems and the resources they provide (ecosystem goods and ecosystem services) as the basic capital for the development of an industrial economy.	• Students read and analyze historical articles and explain why Britain's natural systems and resources provide the basic capital for the development of an industrial economy. • Students identify how the widespread availability of natural resources explains why Britain became the first country to industrialize.
Provide examples of the connection between natural systems and resources, and entrepreneurship, labor, and capital in industrial economies.	• Students analyze information about entrepreneurship, labor, capital, and industrial economies, and write explanations of how each of these factors influenced the development of the Industrial Revolution.
Describe how increased demands provided economic incentives to improve the methods used to extract, harvest, transport, and produce goods from available natural resources.	• Students read historical documents and write an analysis of how demands for resources influenced harvest and transport methods for natural resources. • Students describe how industrialization changed how people use natural resources and natural systems.
Recognize that the growth in human populations and human communities in England placed greater demands on natural systems.	• Students describe how population increases in Great Britain in the 1600s, 1700s, and 1800s helped prompt the Industrial Revolution. • Students identify environmental changes caused by industrialization. • Students describe how the demand for ecosystem goods changed as Britain's population grew and cities expanded.

they provide might be called the basic capital for the development of an industrial economy," and "Give an example of (and describe) a connection between Britain's natural systems and resources and each of the following: entrepreneurship, labor, and capital."

Like the majority of EBE instructional units, Jackson Elementary School's "Water Works" unit focused on interdisciplinary instruction involving English language arts, math, science, history/social science, and systems-related content.[2] Since their learning objectives paralleled the standards, the fifth-grade teachers at Jackson developed an interdisciplinary assessment that allowed them simultaneously to measure their students' growing proficiency with the standards and their knowledge of the local environment. The focus of their investigation was water use and waste on the campus, so as one of the culminating assessments, teachers had the students apply what they had learned to an investigation of plantings and gardens around the campus. They had the students demonstrate their proficiency with solving real-world math problems, for example, by identifying where water was used on the campus and how much water was used in different areas, and then comparing the volumes of water used in those areas. To assess students' English language arts skills, teachers had students create brochures and make presentations to the PTA in which they explained the results of their study and described their proposal for the school to begin to use "water-wise" plants to save water. The teachers assessed students' scientific and environmental knowledge by analyzing the content of their brochures and presentations.

Student assessments are an important element of instructional practice, and the information gained from these assessments is used as one of the several components of the formative evaluation of the program, as well as a significant element of every summative program evaluation.

EVALUATING EBE PROGRAMS FOR CONTINUOUS IMPROVEMENT

All too often evaluation is not taken into consideration until a program is already underway, or it is thought about only at the end of a program, when implementers start to think about how they are going to report program outcomes to administrators and funders. Yet, the greatest benefits can be gained by considering evaluation needs through all stages of planning and implementing an EBE program.

Jackson Elementary School's "Water Works" Unit

Combined Learning Objectives Driving the Assessment of Standards and Systems

SUMMARY OF LEARNING OBJECTIVES	SUMMARY OF LESSON ACTIVITY AND ASSESSMENTS
English language arts	
Use electronic media to gather information.	• Students use the Internet to gather information and discuss pros and cons of water conservation, and create a graphic organizer for their writing assignment.
Write expository pieces with a purpose and explanation, and offer evidence.	• Students use their graphic organizer as the basis for writing an expository article for a brochure that they then publish.
Give presentations on problems/solutions using multimedia and visual displays.	• Students make presentations using visual aids to persuade an audience of their argument.
Mathematics	
Convert among different-sized standard measurement units.	• Students measure water use at different locations on campus and convert results into different units of volume.
Relate volume to the operations of multiplication and addition.	• Students use data from their water use survey to calculate the total water usage on the campus.
Make plots to display a data set of measurements.	• Students plot the data from their survey of campuswide water use and create histograms and bar graphs to show their individual data.
Science	
Identify sources and percentages of freshwater and saltwater on Earth.	• Students name three sources of freshwater and two seas and oceans, and use a map to estimate the coverage over Earth's surface of freshwater and saltwater.
Locate community-based sources of water and describe ways to conserve it.	• Students identify their reservoir on a local map and identify two ways that they could help conserve water.
Conduct a scientific experiment to identify water usage at the school.	• Students design an investigation to determine how much water is used at the school, based on utility bills.

(continues)

SUMMARY OF LEARNING OBJECTIVES	SUMMARY OF LESSON ACTIVITY AND ASSESSMENTS
History/social science	
Describe how past civilizations gained and conserved water.	• Students identify two ways ancient civilizations collected and transported water.
Describe and compare how early civilizations caused water pollution versus how society pollutes today.	• Students discuss the present-day causes of water pollution and describe how they compare to the past.
Understand water's role in transportation across the United States.	• Students identify three rivers that play an important role in the movement of people, raw materials, and finished goods across the United States.
Natural and human social systems	
Students will list one reason why water is a limited resource in three geographic regions: the city of Altadena, the state of California, and the United States.	• Students list the factors that make water a limited resource.
Describe methods for conserving water at school, at home, and in the community.	• Students write a one-paragraph description of three methods of water conservation.
Participate in a water conservation project at Jackson Elementary School.	• Students make posters about the need for water conservation and display them throughout the school.

This "formative" approach to evaluation can help program designers, administrators, and teachers to think about how they can use evaluation data as part of a strategy for ongoing improvement of an EBE program. While summative evaluations play an equally important role, they are more focused on the "end results" that are reported to administrators, funders, and other stakeholders. This section discusses formative evaluation processes, since they can play such an important role in monitoring and improving EBE programs as they are being designed and implemented (the major focus of this book).

As soon as the members of the planning team have established an EBE program's vision, mission, and goals, they should begin discussing evaluation practices so that progress can be adequately monitored. The first step in designing evaluation is reviewing mission and goals statements and identifying the tasks and measurable outcomes that will be monitored during program implementation. These tasks and measurable outcomes, of course, may change throughout the life of a program even as the mission and goals remain stable.

California's EEI, for example, evolved through several different developmental phases, each with its own set of tasks and measurable outcomes. The first phase focused on developing California's Environmental Principles and Concepts. The measurable outcome for this was receiving approval of the draft versions from the appropriate state agencies. Although there was a multitude of subtasks, the second major task was developing a K–12 curriculum that would bring these principles to schools across the state. This included getting approval by the State Board of Education. Since the 2010 approval, the majority of EEI-related tasks have been focused on getting the curriculum used by schools across the state, an effort that is still in process, but is being measured on a regular basis by determining how many schools and districts are implementing the program.

The evaluation information related to each of these phases of the EEI is formative in nature and only provides data related to the implementation process; it does not provide a meaningful picture of progress toward the EEI's mission—expanding K–12 students' knowledge of natural and human social systems and the environmental issues that result from the interactions between them. Evaluating the progress toward this mission requires evidence related to the curriculum's effects on students' knowledge of the environmental principles and concepts and related state content standards—a summative evaluation that will be undertaken after several more years of implementation.

Environment-based education programs must develop evaluation processes and associated data-gathering instruments to meet their specific needs. These evaluations should take into account which data and strategies will be used to gather formative information, and which will be used for summative data collection and reporting.

Over the years, the State Education and Environment Roundtable (SEER) has determined that it is most important to focus formative data gathering on three key areas: instructional materials, instructional methods, and adequacy of support and resources for implementation. These areas are then subdivided into their component parts for evaluation.

Evaluating Instructional Materials

The evaluation of instructional materials examines how materials are designed, written, and structured. Among other considerations, instructional materials should be reviewed from the perspective of: relevance to content standards; coverage of natural and human social systems; extent of interdisciplinary activities; developmental appropriateness and readability of materials; and inclusion of appropriate student assessment instruments.

In developing their EIC Model programs, schools like Seven Generations Charter School, Jackson Elementary School, and Arabia Mountain High School have designed their instructional materials based on the template that SEER has provided them. At the earliest stages of their work, they submit their draft instructional units to SEER's staff for review and recommendations for revisions. Each of these units is reviewed to ensure that it includes all elements of the instructional template: identification of the local environmental context; choices of standards and learning objectives; structuring of the community-based investigations and associated questions; lesson plans, including assessment; and plans for collaboration and environmental service-learning activities. After the teachers have had the opportunity to revise their draft units, they implement them with their students. Then, based on those experiences and reflection with their colleagues, they decide what, if any, changes they want to make before they implement the instructional unit again.

The evaluation of the instructional units in California's EEI curriculum used a field-testing process for the formative evaluation of the materials. Each participating teacher was asked to complete an evaluation instrument after thoroughly implementing the units that had been assigned to him or her. These instruments included a wide range of questions that covered elements such as the adequacy of the unit for teaching the standards, the clarity of procedures, the utility and appropriateness of the student assessment instruments, and the applicability of the instruction and materials to students with different learning styles. It also requested teachers' input on the graphical design of the units.

Both these cases exemplify the appropriate use of formative evaluation data—the information is being used to improve the program materials before they are finalized. This applies to the revision of the instructional materials per se, as well as to the types of changes teachers may make if they adapt the EIC Model template to their particular school's needs.

Evaluating Instructional Methods

In evaluating the instructional methods used in EBE programs, it is important to take into account two major factors: individual readiness and preparedness, and teaching practices. In this instance, readiness encompasses factors like a teacher's willingness to participate, her understanding of how implementation will affect her classroom management and teaching practices, and her comfort level in regard to the cooperation and support she will receive from colleagues and administrators. Teacher preparedness, on the other hand, refers to an educator's knowledge and skills in the use of the specific teaching methods that are involved in any new program.

The State Education and Environment Roundtable uses a set of three different evaluation instruments that are based on the conceptual framework of the Concerns-Based Adoption Model.[3] These instruments measure the depth and quality of implementation of the program and include a "Stages of Concern" questionnaire, an interview-based "Innovation Configuration," and a rubric-based "Self-Evaluation Guide." The data gathered through these instruments provide insight into where teachers and schools are making progress and/or experiencing challenges. Each of the instruments is administered to individual teachers at the schools that are implementing a new EIC Model program.

The "Stages of Concern" questionnaire explores questions related to teachers' feelings about implementing an EBE program, including their level of buy-in and their concerns about implementation. One question, for example, asks teachers to respond to the statement: "I would like to know how this innovation is better than what we have now."[4] Among other things, the results from these questionnaires provide insight into whether the teachers are more concerned about personal aspects—what they will actually be doing to implement the program—or about the effects of the program on their students.

The results that come from the analysis of these questionnaires change as teachers gain experience implementing EBE programs. For example, in the first year, most teachers' concerns revolve around collaborating with other teachers, coordinating their activities with community members and other partners, and having enough time and other resources available. In many cases, too, they may not be entirely comfortable regarding exactly what it is they are supposed to be doing. With greater experience, teachers in the second year of implementation report that they are more concerned about the day-to-day management of instructional activities. As they get even more comfortable implementing EBE programs, their concerns mature to wanting to ensure that they are having a positive impact on the students. This pattern of change has been reported in many studies of the adoption of innovative programs; it is not just associated with EBE implementation.

There are many ways to evaluate how educators are implementing EBE teaching practices. The State Education and Environment Roundtable uses both its self-evaluation rubrics and an EBE-specific "Innovation Configuration" instrument to evaluate teachers' instruction. These two instruments are focused on all of the core practices of EBE, including among others: use of environmental contexts for instruction; integrated interdisciplinary instruction; community-based investigations; and environmental service learning. In the self-evaluation rubrics, for example, teachers are asked to rate and describe how they "use local natural and community surroundings as a context for standards-based instruction," "provide students with opportunities to investigate real-world community problems and issues," and "provide students with opportunities to explore connections between subject area disciplines and among natural and human social systems."[5]

In many instances, schools will also develop their own self-evaluation instruments to monitor their teaching practices. In parallel with their assessment of the development of students' critical-thinking and problem-solving skills, for example, the Seven Generations Charter School teachers developed a self-evaluation instrument—a "teaching strategies checklist"—based on their research on best practices. In this instance, for example, the teachers want to ensure that their instruction is building opportunities for students to compare and contrast, analyze information, make inferences, synthesize solutions to problems, and also evaluate and defend the validity of their ideas. Teachers use this checklist as a reference point with which to monitor themselves and make sure that the learning activities and experiences they are providing students will develop skills in these important areas. (See appendix: "Seven Generations Charter School Critical-Thinking and Problem-Solving Strategies Teaching Checklist," at the end of this chapter.)

Measuring Levels of Support and Resources

While high-quality, educationally effective instruction is at the heart of EBE programs, they will only be successful and sustained over the long term if they have the support of educational leadership, a substantial level of community involvement, and an adequate funding stream. Without this support and the associated resources, even the best instructional program may not get off the ground, survive changes in administration, or endure the next round of budget cuts. Therefore, it is important to monitor a program's critical working relationships and financial resources, even though the majority of evaluation work should always be focused on the instruction. And, the big question that EBE planning and implementation teams should always keep in mind is: "Do we have the support and resources we need to succeed?"

As part of its overall evaluation efforts with EIC Model schools, SEER uses three approaches to evaluating the adequacy of program support and resources: its innovation configuration instruments; self-evaluation rubrics; and interviews with administrators, members of the educational leadership team, community partners, parents, and other participants. These three sources provide the information needed to evaluate the support from educational leaders in schools and districts, the level of community involvement in planning and implementation, and the availability of resources, services, and long-term funding to sustain program activities.

In assessing the adequacy of support from the educational leaders in schools and school districts, it is important to consider the types and depth of involvement they have in the program. There are many ways that these leaders can support a program, so it is important to evaluate if and how they are involved in program planning, participate in a leadership or advisory committee, assist with fund-raising, or provide the implementation team with planning time or other resources, such as access to transportation and study sites. In evaluating the role of educational leadership in an EBE program, SEER examines:

- The school and district administrators' roles in providing the program with ongoing leadership in instructional design, planning, and evaluation;
- The commitment of the school and district administrators to the long-term support and funding of the program;
- The participation of administrators and their support of teachers in their efforts to find funding and build partnerships with the community.

Evaluating the involvement of community partners requires reaching out to the diverse stakeholder groups in the area, including parents and parent/teacher organizations, government officials and agencies, community groups, businesses, and many others. These partners can be extremely important to the success of the program in both the short and long term, because they can often be significant providers of technical expertise, services, resources, and even funding for program activities. Interviews are usually the most effective approach for gathering this information from the participating individuals, committees, businesses, and community groups. The most important topics to discuss with these individuals and organizations include:

- Their involvement in mentoring programs, internships, and service-learning opportunities that offer a variety of community-based experiences;
- The part they play in helping to find necessary program funds, services, and resources, and in developing strategies to sustain program activities;
- Their roles in sharing information about the program's work with the community, providing recognition for students' and teachers' accomplishments, and giving students and teachers opportunities to showcase their successes.

The information gained from all aspects of the formative evaluations provides the basis for continuously improving the program and for providing educators at the school, district, or state level with any additional technical support, professional development, or resources they may need to create a successful and sustainable EBE program.

WORDS OF ENCOURAGEMENT

The examples of environment-based education that are described throughout this book are not a fantasy or a dream—they are vibrant, living programs that engage students and teachers in active learning that has meaning for their daily lives and for their futures. And, they give students from diverse backgrounds an opportunity to become active, contributing members of the global society of the twenty-first century.

These programs have all been developed by educators and community members who are committed to giving their students and children an education that is like no other. The hundreds of teachers and students who have shared the stories of their EBE programs with me over the past eighteen years have taught me everything I know about how to make this happen. Most important, however, is that this is happening every day in classrooms across the country through programs developed by creative and hard-working teachers and administrators, and supported by community members, researchers, university faculty members, businesses, and many others.

I encourage you to use these "lessons learned" to create an EBE program that can engage and intrigue learners and revitalize teachers. Keep in mind a comment about teaching compared to other professions that we hear again and again from educators across the country: "It may be harder, but I wouldn't give it up for anything."

APPENDIX A (CHAPTER 9): SEVEN GENERATIONS CHARTER SCHOOL CRITICAL-THINKING AND PROBLEM-SOLVING RUBRIC

STUDENT PROGRESS	CRITICAL THINKING	PROBLEM SOLVING
WD **Well developed:** Working above grade level for the student's age/grade placement in a specific area.	• Demonstrates an advanced understanding of the system and makes connections to other systems • Defines multiple issues or problems related to the system • Uses previously gathered evidence to formulate and draw inferences regarding the system	• Clearly identifies the problem as well as secondary issues related to the problem • Develops and precisely explains a viable solution to the problem • Evaluates the possible outcomes of the solution, including problems that can arise in the process of implementing the initial solution
DE **Developing as expected:** Developing as expected in a given skill, concept, or behavioral area. This reflects progress that is appropriate for the age/grade placement. May be modified with + or -.	• Demonstrates a clear understanding of the system • Defines an issue or problem related to the system • Uses previously gathered evidence to understand the system	• Clearly identifies the problem • Develops and explains a solution to the problem • Evaluates the possible outcomes of the solution
BD **Beginning to develop:** Beginning to develop a given skill, concept, or behavioral area. With time and experience, the level of understanding and concept development is expected to reach an appropriate level.	• Demonstrates a limited understanding of the system • Identifies a problem related to the system but overlooks core issues • Uses minimal evidence to understand the system	• Identification of a specific problem is not clear • Offers possible solutions, but solutions are not fully developed • States the possible outcome of a solution but does not reflect an evaluation of that solution
NYA **Not yet apparent:** Not yet demonstrating a given skill, concept, or behavioral area. This does not indicate failure, but rather reflects a different rate of development.	• Understanding of the system is not clearly demonstrated • Issue or problem related to the system is not identified • Evidence is not provided to demonstrate an understanding of the system	• A problem is not identified • Possible solutions are not given or are not developed in any way • No evaluation of the possible solution is given

Source: Developed by Seven Generations Charter School teachers, including: Amanda Cossman, Angela Waldraff, Jennifer Hersh, Lisa Fritz, Louise Moyer, Mara Richardson, Katie Pulizzano, and Niki Rothdeutsch.

APPENDIX B (CHAPTER 9): SEVEN GENERATIONS CHARTER SCHOOL CRITICAL-THINKING AND PROBLEM-SOLVING STRATEGIES TEACHING CHECKLIST

CRITICAL-THINKING AND PROBLEM-SOLVING STRATEGIES	EVIDENCE
Knowledge: Recall, name, define, label, list key concepts	Tree identification based on leaf characteristics, parts of a tree, layers of the forest, life cycle of the tree
Comprehension: demonstrate understanding by …	Draw and diagram the parts of a tree; create life cycle of a tree, create leaf field guide, identifying compound and simple leaves; layers assessment; touch box assessment; habitat drawing; positives and negatives of deer flipbook; camouflage drawing (animal in its habitat); bird model; how-to guide on tree products; and final forest assessment
Compare/contrast	Differences in leaf characteristics, life cycle of a tree, layers of the forest, decomposers, producers, consumers, bird beaks and feet, adopted trees
Interpret and translate	Research on tree products and creation of how-to booklet, tree field guides, bird field guides, mushroom field guides, etc.
Organize and describe	Field guides, seed dispersal methods, research on tree products
Illustrate	Camouflage scene, life cycle of a tree, layers of the forest, diagram of the parts of a tree, compound vs. simple leaves, habitats, deer flipbook, producers, consumers, decomposers, forest mural
State main ideas	Draw and diagram the parts of a tree; create life cycle of a tree: create leaf field guide, identifying compound and simple leaves; layers of the forest assessment; touch box assessment; habitat drawing; positives and negatives of deer flipbook; camouflage drawing (animal in its habitat); bird model; how-to guide on tree products; and final forest assessment
Application: solve problems by …	Habitat garden cleanup, forest self-reflection, canopy removal scenario
Apply, identify, choose plan	Habitat garden cleanup
Develop, model, build, and construct methods	Bird models, birdhouses, bird feeders, mycelium terrarium, forest touch box, deforestation/reforestation, edible forest, wildlife-friendly tree ornaments
Analysis of information	Acorn collection and graphing
Inferring	Through daily discussions, we infer with the students (e.g., deforestation/reforestation impacts on forest animals)

Note: The "Evidence" column represents samples from the school's second-grade "Fantastic Forests" unit.

(continues)

CRITICAL-THINKING AND PROBLEM-SOLVING STRATEGIES	EVIDENCE
Analyze data	Acorn graphing, forestry tree growth tables, temperature and rainfall data (PA vs. rain forest), life cycle of a tree, "stage hike"
Collect evidence	Acorn graphing, leaf characteristics, creation of field guides, forest touch box, seed dispersal, etc.
Classify and categorize	Seed dispersal, layers of the forest, tree identification, life cycle of trees
Synthesize solutions to problems by …	Through daily discussions (e.g., habitat garden cleanup)
Create, modify, design	Bird models, forest mural, camouflage drawings, touch box
Test, predict, improve	Habitat garden, mycelium growth rates
Evaluation—present and defend validity of ideas	Habitat garden (needs)
Self-reflect and evaluate	Forest self-reflection
Measure and defend results	Mycelium growth rate project
Systems thinking—analyze how parts interact to produce overall outcomes	Layers of the forest, deforestation/reforestation, products that come from trees, stages of the tree lifecycle, biodiversity
Synthesize and make connections between information and arguments	Mycelium growth project, deforestation/reforestation, sap collection project
Solve different kinds of unfamiliar problems in conventional and unconventional ways	Health of South Mountain Forest (trash walk/habitat garden)
Identify and ask questions that clarify points of view and lead to better solutions	Forest self-reflection, habitat garden
Divergent questioning	Forest self-reflection, habitat garden
Research process	Bird models, products from trees, tree field guides
Scientific method	Tree field guides, mycelium growth project, soil exploration, mold experiment
Metacognitive reflection	Forest self-reflection

Source: Developed by Seven Generations Charter School teachers, including: Amanda Cossman, Angela Waldraff, Jennifer Hersh, Lisa Fritz, Louise Moyer, Mara Richardson, Katie Pulizzano, and Niki Rothdeutsch.

California's Environmental Principles and Concepts

The environmental principles examine the interactions and interdependence of human societies and natural systems. The nature of these interactions is summarized in the environmental principles and concepts that follow.[1]

PRINCIPLE I

The continuation and health of individual human lives and of human communities and societies depend on the health of the natural systems that provide essential goods and ecosystem services. As a basis for understanding this principle:

Concept a. Students need to know that the goods produced by natural systems are essential to human life and to the functioning of our economies and cultures.

Concept b. Students need to know that the ecosystem services provided by natural systems are essential to human life and to the functioning of our economies and cultures.

Concept c. Students need to know that the quality, quantity, and reliability of the goods and ecosystem services provided by natural systems are directly affected by the health of those systems.

PRINCIPLE II

The long-term functioning and health of terrestrial, freshwater, coastal, and marine ecosystems are influenced by their relationships with human societies. As a basis for understanding this principle:

Concept a. Students need to know that direct and indirect changes to natural systems due to the growth of human populations and their consumption rates influence the geographic extent, composition, biological diversity, and viability of natural systems.

Concept b. Students need to know that methods used to extract, harvest, transport, and consume natural resources influence the geographic extent, composition, biological diversity, and viability of natural systems.

Concept c. Students need to know that the expansion and operation of human communities influences the geographic extent, composition, biological diversity, and viability of natural systems.

Concept d. Students need to know that the legal, economic, and political systems that govern the use and management of natural systems directly influence the geographic extent, composition, biological diversity, and viability of natural systems.

PRINCIPLE III

Natural systems proceed through cycles that humans depend upon, benefit from, and can alter. As a basis for understanding this principle:

Concept a. Students need to know that natural systems proceed through cycles and processes that are required for their functioning.

Concept b. Students need to know that human practices depend upon and benefit from the cycles and processes that operate within natural systems.

Concept c. Students need to know that human practices can alter the cycles and processes that operate within natural systems.

PRINCIPLE IV

The exchange of matter between natural systems and human societies affects the long-term functioning of both. As a basis for understanding this principle:

Concept a. Students need to know that the effects of human activities on natural systems are directly related to the quantities of resources consumed and to the quantity and characteristics of the resulting by-products.

Concept b. Students need to know that the by-products of human activity are not readily prevented from entering natural systems and may be beneficial, neutral, or detrimental in their effect.

Concept c. Students need to know that the capacity of natural systems to adjust to human-caused alterations depends on the nature of the system as well as the scope, scale, and duration of the activity and the nature of its by-products.

PRINCIPLE V

Decisions affecting resources and natural systems are based on a wide range of considerations and decision-making processes. As a basis for understanding this principle:

Concept a. Students need to know the spectrum of what is considered in making decisions about resources and natural systems and how those factors influence decisions.

Concept b. Students need to know the process of making decisions about resources and natural systems, and how the assessment of social, economic, political, and environmental factors has changed over time.

Summary of the EBE Research Data

The following lists summarize the research data related to EBE programs that is currently available. The *n* for grade-level student assessment comparisons indicates the number of separate assessments that were used to compare "treatment" and "control" groups of students, it is not an indicator of the total number of students participating in each assessment.

ENGLISH LANGUAGE ARTS (ELA)

Summary of Results

- 100% of standardized test scores indicated that students in EIC Model programs performed better than students in traditional programs (*n* = 17 grade-level student assessment comparisons in nine schools in five states).[1]
- 76% of standardized test scores indicated that students in the EBE programs performed better than students in traditional programs (*n* = 91 grade-level student assessment comparisons in eight pairs of California schools).[2]
- 98% of standardized test scores indicated that students in the EBE programs performed as well or better than students in traditional programs (*n* = 240 grade-level student assessment comparisons in four pairs of California schools with a total of 3,720 students).[3]
- Students in 66% of schools with EBE programs scored higher (statistically significant) on Washington Assessment of Student Learning (WASL) reading tests than otherwise comparable schools over a five-year study period. The same

patterns of higher scores appeared in WASL assessments of writing (73% of schools) and listening (60% of schools) ($n = 77$ pairs of Washington schools).[4]

- 80% of grade-level comparisons demonstrated higher scores on Palmetto Achievement Challenge Tests (PACT) by students in the EIC Model programs ($n = 10$ pairs of classrooms in ten South Carolina schools).[5]
- Higher rubric-based writing scores (statistically significant) following environment-based lessons designed to teach a state writing standard ($n = 2$ teachers in two Florida schools).[6]
- 9% increase in the number of students performing at the satisfactory level on LEAP 21 in Louisiana's East Feliciana School District over a four-year period after implementing a place-based program.[7]
- 5% higher score at school 1, after implementing of an EIC Model program, on Florida Comprehensive Assessment Test (FCAT) reading assessment over fourth-grade students from the previous year.[8]
- 3% higher score at school 2, after implementing an EIC Model program, on Florida Comprehensive Assessment Test (FCAT) reading assessment over fourth-grade students from the previous year.[9]
- 30% higher score at school 3, after implementing an EIC Model program, on *Florida Writes!* assessment over fourth-grade students from the previous year.[10]
- 7% higher score at school 2, after implementing an EIC Model program, on *Florida Writes!* assessment over fourth-grade students from the previous year.[11]
- 14% higher score at school 4, after implementing an EIC Model program, on *Florida Writes!* assessment over eighth-grade students from the previous year.[12]
- Students in the first-grade EBE program consistently scored higher on Iowa Tests of Basic Skills (ITBS) reading, listening, language, and word analysis assessments than all other first-grade students in the school (students were not grouped into classes by their abilities) ($n = 1$ Texas school).[13]
- 9% increase in reading proficiency of students in third-, fourth-, and fifth-grade classes after the EBE program was implemented, a change from below to above the state average in fourth and fifth grades ($n = 1$ North Carolina school).[14]
- 11% increase in writing proficiency of students in third-, fourth-, and fifth-grade classes after the EBE program was implemented, a change from below to above the state average ($n = 1$ North Carolina school).[15]

- 83% of students in the EBE program scored at or above the "proficient" level on the Wisconsin reading assessment, compared with only 38% in all other Wisconsin schools at the same income level ($n = 1$ Wisconsin school).[16]
- 44% increase in Kentucky Instructional Results Information System (KIRIS) reading scores of fourth-grade students during the four years after the EBE program was initiated ($n = 1$ Kentucky school).[17]
- 33% average increase, after implementing the EBE program, on *Florida Writes!* narrative writing assessment over a five-year study of fourth-grade students ($n = 5$ Florida schools).[18]
- 47% average increase, after implementing the EBE program, on *Florida Writes!* expository writing assessment over a five-year study of fourth-grade students ($n = 5$ Florida schools).[19]
- 9% average increase, after implementing the EBE program, on Florida Comprehensive Assessment Test (FCAT) reading over a five-year study of fourth-grade students ($n = 5$ Florida schools).[20]
- 12% higher scores for students on Colorado Student Assessment Program (CSAP) reading assessment in schools implementing an EBE program over students in traditional programs within a four-county comparison group of schools ($n = 317$ students in program compared to weighted-average assessment scores from four feeder school districts).[21]
- 4% higher scores for students on Colorado Student Assessment Program (CSAP) writing assessment in schools implementing an EBE program over students in traditional programs within a four-county comparison group of schools ($n = 317$ students in program compared to weighted-average assessment scores from four feeder school districts).[22]
- 11% more students in EBE (Integrated) program meeting Washington state standards on reading compared to students in traditional program (seven-year study).[23]
- 11.5% more students in EBE (Integrated) program meeting Washington state standards on writing compared to students in traditional program (seven-year study).[24]
- Fourth-grade students in an EIC school outperformed (statistically significant) students on language arts when compared to comparable students in a non-EIC school.[25]

MATH

Summary of Results

- 71% of standardized test scores indicated that students in EIC Model programs performed better than students in traditional programs ($n = 7$ grade-level student assessment comparisons in seven schools in five states).[26]

- 63% of standardized test scores indicated that students in the EBE programs performed better than students in traditional programs ($n = 27$ grade-level student assessment comparisons in eight pairs of California schools).[27]

- 92% of standardized test scores indicated that students in the EBE programs performed as well or better than students in traditional programs ($n = 80$ grade-level student assessment comparisons in four pairs of California schools with a total of 3,720 students).[28]

- Students in 65% of schools with EBE programs scored higher (statistically significant) on Washington Assessment of Student Learning (WASL) math tests than otherwise comparable schools over a five-year study period ($n = 77$ pairs of Washington schools).[29]

- 70% of grade-level comparisons demonstrated higher scores on Palmetto Achievement Challenge Tests (PACT) by students in the EIC Model programs ($n = 10$ pairs of classrooms in ten South Carolina schools).[30]

- 4% increase in number of Alaska native students scoring "proficient" on Alaska state benchmark examination for eighth graders in schools implementing place-based programs through the Alaska Rural Systemic Initiative (AKRSI) vs. non-AKRSI schools.[31]

- 5.6 point differential (increase) from tenth-grade Alaska native students in schools implementing place-based programs through the AKRSI vs. non-AKRSI schools.[32]

- Significantly higher rate of movement of eleventh-grade Alaska native students from out of the lower quartile in schools implementing place-based programs through the AKRSI vs. non-AKRSI schools. However, students in the non-AKRSI schools entered the top quartile at a faster pace than students in the AKRSI schools.[33]

- 19% increase in the number of students performing at the satisfactory level on LEAP 21 in Louisiana's East Feliciana School District over a four-year period after implementing a place-based program.[34]

- 6% higher score, after implementing an EIC Model program, on Florida Comprehensive Assessment Test (FCAT) mathematics assessment over fourth-grade students from the previous year.[35]
- Texas students in the first-grade EBE program consistently scored higher on Iowa Tests of Basic Skills (ITBS) math concepts, advanced problem skills, and computation assessments than all other first-grade students in the school (students were not grouped into classes by their abilities) ($n = 1$ Texas school).[36]
- 6% increase in math proficiency of students in third-, fourth-, and fifth-grade classes after the EBE program was implemented, a change from below to above the state average in fourth and fifth grades ($n = 1$ North Carolina school).[37]
- 48% of students in the EBE program scored at or above the "proficient" level on the Wisconsin reading assessment compared with only 15% in all other Wisconsin schools at the same income level ($n = 1$ Wisconsin school).[38]
- 51% average increase, after implementing the EBE program, on Florida Comprehensive Assessment Test (FCAT) math over a five-year study of fourth-grade students ($n = 5$ Florida schools).[39]
- 35% increase on Massachusetts Comprehensive Assessment System (MCAS) test scores for eighth-grade students at Beebe School after implementing a place-based program. Students scored higher in math than state and district averages.[40]
- 10% more students in EBE (Integrated) program meeting Washington state standards on math compared to students in traditional program (seven-year study).[41]
- Fourth-grade students in an EIC school outperformed (statistically significant) students on math when compared to comparable students in a non-EIC school.[42]

SCIENCE

Summary of Results

- 75% of standardized test scores indicated that students in EIC Model programs performed better than students in traditional programs ($n = 4$ grade-level student assessment comparisons in three schools in three states).[43]

- 64% of standardized test scores indicated that students in the EBE programs performed better than students in traditional programs ($n = 11$ grade-level student assessment comparisons in eight pairs of California schools).[44]
- 90% of grade-level comparisons demonstrated higher scores on Palmetto Achievement Challenge Tests (PACT) by students in the EIC Model programs ($n = 10$ pairs of classrooms in ten South Carolina schools).[45]
- 8% increase in the number of students performing at the satisfactory level on LEAP 21 in Louisiana's East Feliciana School District over a three-year period after implementing a place-based program.[46]
- 108% increase in Kentucky Instructional Results Information System (KIRIS) science scores of fourth-grade students during the four years after the EBE program was initiated ($n = 1$ Kentucky school).[47]
- 22% increase on Massachusetts Comprehensive Assessment System (MCAS) test scores for eighth-grade students at Beebe School after implementing a place-based program. Students scored higher in the life sciences than state and district averages.[48]
- 6% higher scores for students on Colorado Student Assessment Program (CSAP) science assessment in schools implementing an EBE program over students in traditional programs within a four-county comparison group of schools ($n = 317$ students in program compared to weighted-average assessment scores from four feeder school districts).[49]
- 8% more students in EBE (Integrated) program meeting Washington state standards on science compared to students in traditional program (five-year study).[50]
- Fourth-grade students in an EIC school outperformed (statistically significant) students on science when compared to comparable students in a non-EIC school.[51]

HISTORY/SOCIAL SCIENCE

Summary of Results

- 100% of standardized test scores indicated that students in EIC Model programs performed better than students in traditional programs ($n = 2$ grade-level student assessment comparisons in two schools in two states).[52]

- 73% of standardized test scores indicated that students in the EBE programs performed better than students in traditional programs ($n = 11$ grade-level student assessment comparisons in eight pairs of California schools).[53]
- 70% of grade-level comparisons demonstrated higher scores on Palmetto Achievement Challenge Tests (PACT) by students in the EIC Model programs ($n = 10$ pairs of classrooms in ten South Carolina schools).[54]
- 11% increase in the number of students performing at the satisfactory level on LEAP 21 in Louisiana's East Feliciana School District over a three-year period after implementing a place-based program.[55]
- 133% increase in Kentucky Instructional Results Information System (KIRIS) social studies scores of fourth-grade students during the four years after the EBE program was initiated ($n = 1$ Kentucky school).[56]
- Fourth-grade students in an EIC school outperformed (statistically significant) students on social studies when compared to comparable students in a non-EIC school.[57]

DISCIPLINARY ACTIONS, ATTENDANCE, AND STUDENT ATTITUDES

Summary of Results

- 100% of assessments of student behavior indicated that students in EIC Model programs performed better than students in traditional programs ($n = 4$ grade-level comparisons).[58]
- 8% fewer "discipline interactions" occurred among students in EBE program (Integrated) than those in the traditional program.[59]
- Students in EBE program (Integrated) scored 11 percentile points higher than traditional-program peers on the SAM (School Attitude Measure).[60]
- 100% of assessments of student attendance and attitudes indicated that students in EIC Model programs performed better than students in traditional programs ($n = 5$ grade-level comparisons).[61]
- 77% of assessments of student attendance indicated that students in the EBE programs performed better than students in traditional programs ($n = 22$ grade-level comparisons in eight pairs of California schools).[62]

- Middle school students in South Carolina's EIC Model program were 78% less likely to be the subject of in-school disciplinary actions than their peers in traditional school programs (n = six South Carolina schools).[63]
- Higher scores (statistically significant) on an assessment of "students' achievement motivation" for both ninth- and twelfth-grade students (n = 400 students in eleven Florida schools).[64]
- 10.8% lower tardiness rate for students in the EIC Model program compared to students in traditional program.[65]
- 29% lower unauthorized absence rate for students in the EIC Model program compared to students in traditional program.[66]
- 5.7% lower insubordination rate for students in the EIC Model program compared to students in traditional program.[67]

CRITICAL THINKING

Summary of Results

- 96% of teachers reported that their students were more proficient in critical thinking as a result of implementing their EIC Model program (n = 167 teachers in forty schools in thirteen states).[68]
- 97% of teachers reported that their students were more proficient in solving problems and thinking strategically as a result of implementing their EIC Model program (n = 167 teachers in forty schools in thirteen states).[69]
- 96% of teachers reported that their students were more proficient in decision making as a result of implementing their EIC Model program (n = 168 teachers in forty schools in thirteen states).[70]
- Critical thinking skills of ninth-grade students, as measured by the Cornell Critical Thinking Test, were higher (statistically significant) than the control group after implementing an EBE program (n = 400 students in eleven Florida schools).[71]
- Critical thinking skills of twelfth-grade students, as measured by the Cornell Critical Thinking Test, were higher (statistically significant) than the control group. Likewise, the "disposition toward critical thinking" of twelfth-grade students, as measured by the California Measure of Mental Motivation, were

higher (statistically significant) than the control group after implementing an EBE program (n = 400 students in eleven Florida schools).[72]

- Ninth- and twelfth-grade students in EBE programs had critical thinking skill levels comparable to or exceeding those of college students in several American universities, as evidenced by test norms, after implementing an EBE program (n = 400 students in eleven Florida schools).[73]

Types of Support That Stakeholders and Partners Can Provide

The following lists provide examples of the types of support and assistance that stakeholders and partners can provide during the planning and implementation of an EBE program.

Participating in program planning:

- Drafting program plans
- Advising about local environmental and community issues
- Giving insights into parental interests and concerns
- Sharing knowledge about key decision makers in the community
- Providing insights into budgeting and administrative matters
- Advising about potential sources of funding and material resources
- Offering knowledge of local, district, and state educational policies and requirements

Providing technical knowledge and professional expertise:

- Supporting both teachers and students with technical information related to subject matter in which they lack or have only limited expertise (environmental issues, community history, scientific knowledge, etc.)
- Offering different perspectives on local environmental and community issues
- Providing knowledgeable speakers to support specific lessons and content
- Introducing students to career and college opportunities

Supplying implementation resources and financial support:

- Providing materials or equipment for student projects such as habitat restoration
- Fund-raising and/or directly providing funds for instructional materials, supplies, or items such as transportation
- Writing grant proposals

Developing community and political support:

- Championing the program with educational administrators, parent-teacher organizations, and school board members in order to garner support or encourage the development of policies that allow the implementation of an EBE program
- Communicating, networking, and coordinating with other community members, technical experts, and funding sources to support the teachers' and school administrators' efforts
- Organizing or sponsoring and supporting community events
- Advising on matters that may be sensitive to the community culturally, economically, or politically
- Giving students a sense of belonging to their community

Offering learning opportunities and instructional support:

- Participating in student instruction in the classroom, leading instructional field trips, getting their "hands dirty" along with students when they are involved in projects outside of the classroom, such as restoring habitat, building trails, and constructing structures
- Challenging students and teachers to explore new ideas about the environment and their local community
- Providing locations where students can get "on-the-ground" experiences, as well as undertaking investigations and other projects
- Assisting with group management when students and teachers are involved in activities inside and outside of the classroom setting

APPENDIX D

Examples of EBE Community Partners

CALIFORNIA'S EDUCATION AND THE ENVIRONMENT INITIATIVE

State Government Partners

- State Assemblymember Fran Pavley's office
- California Department of Education
- California Environmental Protection Agency
- California Integrated Waste Management Board
- California Natural Resources Agency
- Governor's Office of the Secretary of Education
- State Senator Tom Torlakson's office
- State Board of Education

External Partners

- California Council for Environmental and Economic Balance (business/industry association)
- California Farm Bureau Federation (business/industry association)
- California Film Extruders & Converters Association (industry association)
- California Forest Products Commission (business/industry association)
- California Institute for Biodiversity (nongovernmental environmental organization)
- California Manufacturers & Technology Association (business/industry association)
- California School Boards Association (professional association)

- California Science Teachers Association (professional association)
- California State Association of Counties (professional association)
- California State Parks Foundation (nongovernmental environmental organization)
- California State PTA (professional association)
- California Teachers Association (professional association)
- Coalition for Clean Air (nongovernmental environmental organization)
- EMA, Inc. (environmental consulting business)
- George Lucas Educational Foundation (nonprofit foundation)
- Heal the Bay (nongovernmental environmental organization)
- League of California Cities (professional association)
- National Geographic Society (nongovernmental educational organization)
- Planning and Conservation League (nongovernmental environmental organization)
- Scripps Institution of Oceanography (university research institution)
- Sierra Club Youth Services (nongovernmental environmental organization)
- Society of the Plastics Industry, Flexible Film & Bag Division (industry association)
- TreePeople (nongovernmental environmental organization)
- Universal Studios (business/industry)
- University of California, Berkeley, School of Public Health
- University of Southern California Sea Grant
- The Walt Disney Company (business/industry)
- Warner Brothers (business/industry)
- Waste Management, Inc. (business/industry)
- Water Education Foundation (nongovernmental environmental organization)
- Western States Petroleum Association (business/industry association)
- Wright Consulting (educational consulting business)

DESERT SANDS UNIFIED SCHOOL DISTRICT, LA QUINTA, CALIFORNIA

- Big Morongo Canyon Preserve
- Coachella Valley Preserve
- Coachella Valley Wild Bird Center

- Coachella Valley Women for Agriculture
- Joshua Tree National Park
- Mother Nature Live
- Palm Springs Art Museum
- PEAK, The Energy Coalition
- Santa Rosa and San Jacinto Mountains National Monument
- The Living Desert
- University of California, Berkeley
- University of Redlands

JACKSON ELEMENTARY SCHOOL, ALTADENA, CALIFORNIA

- Altadena Town Council
- Jackson Elementary School facilities management staff
- Jackson Elementary School Parent Teacher Association
- Land Use Committee of the Altadena Town Council
- Metropolitan Water District of Southern California
- Pasadena Educational Foundation
- Pasadena Unified School District (science coordinator, purchasing and accounting offices, facilities management staff)
- Various local businesses
- Various local environmental groups

SEVEN GENERATIONS CHARTER SCHOOL, EMMAUS, PENNSYLVANIA

- 1803 House, Emmaus
- Bethlehem Landfill
- Cedar Crest College
- Cougles Recycling
- Dan Schantz Greenhouse & Cut Flower Outlet
- DaVinci Science Center
- Delaware Canal State Park
- Delaware River Shad Fishermen's Association
- Edge of the Woods Nursery

- Emmaus Borough Compost Center
- Emmaus Borough manager
- Emmaus Borough Public Works
- Emmaus Borough Water Department
- Emmaus Historical Society
- Emmaus Public Library
- Ewe Can Do It Shetland Sheep Farm
- Frecon Farms
- Frick Family Farm
- Good Shepherd Rehabilitation
- Gottschell Farm
- Herbines Garden Center
- Historic Bethlehem Partnership
- Hunter Hill Farms
- HydroMania
- Independent composting expert
- Independent native plant specialist
- Independent soil scientist
- Independent wetland consultant
- Jacobsburg Environmental Education Center
- Jaindl Farms
- Knauss Homestead
- Lehigh County Historical Society
- Lehigh Valley Zoo
- Lost River Caverns
- Marwell Dairy Farm
- Monroe County Conservation District
- Muhlenberg College
- Pennsylvania Department of Conservation and Natural Resources
- Pennsylvania Department of Labor & Industry
- Pennsylvania Fish and Boat Commission
- Pennsylvania German Cultural Heritage Center
- Pennsylvania state representative
- Rodale Press

- Scranton Museum of Transportation
- Shelter House
- Unangst Tree Farms
- U.S. representative (retired)
- Vista Farm
- Waste Management, Inc.
- Wildlands Conservancy

Notes

Introduction

1. Jennifer McMurrer, *Instructional Time in Elementary Schools: A Closer Look at Changes for Specific Subjects* (Washington, DC: Center on Educational Policy, 2008); Noelle Ellerson, *Weathering the Storm: How the Economic Recession Continues to Impact School Districts* (Alexandria, VA: American Association of School Administrators, March 2012).
2. Ellerson, *Weathering the Storm.*
3. United Nations Educational, Scientific and Cultural Organization, *Intergovernmental Conference on Environmental Education: Final Report* (Paris: UNESCO, 1978), http://unesdoc.unesco.org/images/0003/000327/032763eo.pdf.
4. Kevin Coyle, *Environmental Literacy in America: What Ten Years of NEETF/Roper Research and Related Studies Say About Environmental Literacy in the U.S.* (Washington, DC: The National Environmental Education and Training Foundation, 2005).
5. McMurrer, *Instructional Time in Elementary Schools.*

Chapter 1

1. National Research Council, *National Science Education Standards* (Washington, DC: National Academy Press, 1996), 116.
2. U.S. Geological Survey, "Pig Iron Statistics December 2010," in *Historical Statistics for Mineral and Material Commodities in the United States: U.S. Geological Survey Data Series 140,* comps. T. D. Kelly and G. R. Matos, http://minerals.usgs.gov/ds/2005/140/ironsteel.pdf.
3. Richard Leete et al., *Global Population and Water: Access and Sustainability* (New York: United Nations Population Fund, 2003), 55.
4. American Association of School Administrators, "A Childhood Epidemic," *School Governance & Leadership* 5, no. 1 (Spring 2003): 4.
5. Edgar G. Hertwich et al., *Assessing the Environmental Impacts of Consumption and Production: Priority Products and Materials, A Report of the Working Group on the Environmental Impacts of Products and Materials to the International Panel for Sustainable Resource Management* (Nairobi: United Nations Environment Programme, 2010), 108.

6. U.S. Energy Information Administration, "What Are Greenhouse Gases?" http://www.eia.gov/oiaf/1605/ggccebro/chapter1.html.

7. China Daily Information Co., "Experts Explore Beijing's Air Pollution," http://www.chinadaily.com.cn/china/2012-06/11/content_15494160.htm.

8. U.S. Environmental Protection Agency, "U.S. and Global Temperature," http://www.epa.gov/climatechange/pdfs/print_temperature-2012.pdf.

9. Brian Kahn, "Superstorm Sandy and Sea Level Rise," *Climate Watch Magazine,* November 5, 2012, http://www.climatewatch.noaa.gov/article/2012/superstorm-sandy-and-sea-level-rise.

10. Richard J. Reiner, "Protecting Biodiversity on Grazed Grasslands in California," http://californiarangeland.ucdavis.edu/Publications%20pdf/CRCC/Protecting%20Biodiversity%20on%20Grazed%20Grasslands%20in%20California.pdf.

11. Francesca Grifo and Joshua Rosenthal, eds., *Biodiversity and Human Health: A Guide for Policymakers* (New York: Center for Biodiversity and Conservation, American Museum of Natural History, 1997), http://www.amnh.org/our-research/center-for-biodiversity-conservation/publications/for-policymakers/biodiversity-and-human-health-a-guide-for-policymakers.

12. UN-Water, "Statistics: Graphs & Maps," http://www.unwater.org/statistics.html.

13. Emily Corcoran et al., eds., *Sick Water? The Central Role of Wastewater Management in Sustainable Development.* A Rapid Response Assessment. United Nations Environment Programme, UN-HABITAT, GRID-Arendal, www.grida.no/publications/rr/sickwater.

14. FAO Fisheries and Aquaculture Department, *The State of World Fisheries and Aquaculture 2010* (Rome: Food and Agriculture Organization of the United Nations, 2010), 197.

15. Working Group on Re-Evaluation of Biological Reference Points for New England Groundfish, *Re-Evaluation of Biological Reference Points for New England Groundfish* (Woods Hole, MA: Northeast Fisheries Science Center, 2002), 232.

16. Victor S. Kennedy et al., *Coastal and Marine Ecosystems & Global Climate Change* (Washington, DC: Pew Center on Global Climate Change, 2002), 51.

17. NOAA Marine Debris Program, "What We Know About: The 'Garbage Patches,'" http://marinedebris.noaa.gov/info/patch.html.

Chapter 2

1. Richard Louv, *Last Child in the Woods: Saving Our Children from Nature Deficit Disorder* (Chapel Hill, NC: Algonquin Books, 2006).

2. Liberty H. Bailey, *The Nature-Study Idea, Being an Interpretation of the New School Movement to Put the Child in Sympathy with Nature* (New York: Doubleday, Page & Company, 1903), 10.

3. Ibid., 4.

4. U.S. Forest Service, *Conservation Education in the Forest Service* (Washington DC: U.S. Forest Service, 1999), http://www.fs.usda.gov/Internet/FSE_DOCUMENTS/fsmrs_100523.pdf.

5. Antioch New England Institute, "The Center for Place-Based Education," http://www.antiochne.edu/anei/cpbe/.

6. Committee to Review and Assess the Health and Productivity Benefits of Green Schools, National Research Council, *Green Schools: Attributes for Health and Learning* (Washington DC: The National Academies Press, 2006), http://books.nap.edu/catalog.php?record_id=11756#toc.

7. The College Board, *AP Environmental Science Course Description* (New York: The College Board, 2010), 4, www.apcentral.collegeboard.com/apc/public/repository/ap-environmental-science-course-description.pdf.

8. Gerald A. Lieberman, *Pieces of a Puzzle: An Overview of the Status of Environmental Education in the United States* (Poway, CA: Science Wizards, 1995).

9. United Nations Educational, Scientific and Cultural Organization, *Intergovernmental Conference on Environmental Education: Final Report* (Paris: UNESCO, 1978), 25, http://unesdoc.unesco.org/images/0003/000327/032763eo.pdf.

10. Linda L. Hoody, *The Educational Efficacy of Environmental Education* (San Diego: State Education and Environment Roundtable, 1995), http://www.seer.org/pages/research/educeff.pdf.

11. Gerald A. Lieberman and Linda L. Hoody, *Closing the Achievement Gap: Using the Environment as an Integrating Context for Learning* (San Diego: State Education and Environment Roundtable, 1998), i.

12. The National Service-Learning Clearinghouse defines service learning: "Service-Learning is a teaching and learning strategy that integrates meaningful community service with instruction and reflection to enrich the learning experience, teach civic responsibility, and strengthen communities."

13. Achieve, Inc., *Next Generation Science Standards* (Washington, DC: Achieve, Inc., 2013), 9.

14. Ibid., 93.

Chapter 3

1. Gerald A. Lieberman and Linda L. Hoody, *Closing the Achievement Gap: Using the Environment as an Integrating Context for Learning* (San Diego: State Education and Environment Roundtable, 1998).

2. Gerald A. Lieberman et al., *California Student Assessment Project, Phase 1: The Effect of Environment-Based Education on Student Achievement* (San Diego: State Education and Environment Roundtable, 2000), 20.

3. Edward Falco, South Carolina Department of Education, e-mail message to author with unpublished report "South Carolina 2004–05 PACT Scores, EIC vs. non-EIC Students," March 14, 2006.

4. Tahoma High School, "Endeavor: An Integrated Learning Model for High School Students" (Maple Valley, WA: Tahoma High School, 2008), 22, www.leadered.com/msc08/PowerPointsHandouts/Tahoma MSC PPT.ppt.

5. Tahoma High School, "Endeavor," 23.

6. *The Quality Schools of Choice: Ideas for Instructional Innovation* (New York: McGraw-Hill School Publishing Company and Business Week, 1993), 5.

7. O. Bartosh, "Environmental Education: Improving Student Achievement" (master's thesis, The Evergreen State College, 2003), http://www.seer.org/pages/research/Bartosh2003.pdf.

8. Falco, "South Carolina 2004–05 PACT Scores."

9. Tahoma High School, "Endeavor," 24.

10. Falco, "South Carolina 2004–05 PACT Scores."

11. Allan Sterbinsky, *Rocky Mountain School of Expeditionary Learning Evaluation Report* (Memphis: Center for Research in Educational Policy, University of Memphis, 2002), 10.

12. Lieberman et al., *California Student Assessment Project, Phase 1.*

13. Emeka Emekauwa, *They Remember What They Touch . . . The Impact of Place-Based Learning in East Feliciana Parish* (Washington, DC: The Rural School and Community Trust, 2004), 5.

14. Falco, "South Carolina 2004–05 PACT Scores."

15. Edward Falco, *Environment-Based Education: Improving Attitudes and Academics for Adolescents* (Columbia, South Carolina: South Carolina Department of Education, 2004).

16. Kathy Abrams, "Summary of Project Outcomes from EE & SSS Schools' Final Report Data" (unpublished report, Office of Environmental Education, Tallahassee, FL, 1999), 3.

17. Gerald A. Lieberman et al., *California Student Assessment Project, Phase 1*, 20.

18. Abrams, "Summary of Project Outcomes."

19. Ibid.

20. Julie Athman and Martha Monroe, "The Effects of Environment-Based Education on Students' Achievement Motivation," *Journal of Interpretation Research* 9, no. 1 (2004): 9–25.

Chapter 4

1. LEED certification is granted by the U.S. Green Building Council in recognition of construction that follows a variety of "green" building practices in regard to issues such as selection of construction materials, environmental site assessments, and children's health issues.

2. Gerald A. Lieberman and Jennifer Rigby, *Education and the Environment Initiative: Model Curriculum Plan* (Sacramento: State of California, 2005), http://www.seer.org/EEI_Sources/EEI_MCP.pdf.

3. Gerald A. Lieberman and Linda L. Hoody, *Closing the Achievement Gap: Using the Environment as an Integrating Context for Learning* (San Diego: State Education and Environment Roundtable, 1998).

Chapter 5

1. Environmental service-learning activities are projects that encompass all of the attributes of high-quality service learning and are focused on resolving environmental issues.

2. Based on materials originally developed by Lisa Fritz, Amanda Cossman, and Louise Moyer, "Fantastic Forests" (unpublished, Seven Generations Charter School, Emmaus, PA, 2013), 34.

3. Pennsylvania Department of Education, *Academic Standards for Reading, Writing, Speaking, and Listening* (Harrisburg: Pennsylvania Department of Education, 2012), 7, 10.

4. Common Core State Standards Initiative, *Common Core State Standards for English Language Arts & Literacy in History/Social Studies, Science, and Technical Subjects* (Washington, DC: National Governors Association Center for Best Practices and the Council of Chief State School Officers, 2012), 11, 13, 19.

5. Pennsylvania Department of Education, *Academic Standards for Mathematics* (Harrisburg: Pennsylvania Department of Education, 2010), 8, 11.

6. Common Core State Standards Initiative, *Common Core State Standards for Mathematics* (Washington, DC: National Governors Association Center for Best Practices and the Council of Chief State School Officers, 2012), 19, 20.

7. Pennsylvania Department of Education, *Academic Standards for Science and Technology and Engineering Education* (Harrisburg: Pennsylvania Department of Education, 2012), 9.

8. Pennsylvania Department of Education, *Academic Standards for Environment and Ecology* (Harrisburg: Pennsylvania Department of Education, 2012), 13.

9. Pennsylvania Department of Education, *Academic Standards for Geography* (Harrisburg: Pennsylvania Department of Education, 2012), 6, 7.

10. Based on materials originally developed by Erica Peters and Dennis Silverio, "Water Works" (unpublished, Jackson Elementary School, Altadena, CA, 2008), 40.

11. The author converted the original California English Language Arts and Mathematics standards to the language of the comparable Common Core State Standards. California Department of Education, *Common Core State Standards for English Language Arts & Literacy in History/Social Studies, Science, and Technical Subjects* (Sacramento: California Department of Education, 2013), 9, 17, 25.

12. California Department of Education, *Science Content Standards for California Public Schools* (Sacramento: California Department of Education, 1998), 16.

13. California Department of Education, *History/Social Science Content Standards for California Public Schools* (Sacramento: California Department of Education, 1998), 16, 20.

Chapter 6

1. James A. Farmer Jr., et al., "Cognitive Apprenticeship: Implications for Continuing Professional Education," *New Directions in Adult and Continuing Education* 55 (1992): 41.

2. Based on materials originally developed by Erica Peters and Dennis Silverio, "Water Works" (unpublished, Jackson Elementary School, Altadena, CA, 2008), 40.

3. Based on materials originally developed by Charles P. Hammer and Lori D. Mann, *Britain Solves a Problem and Creates the Industrial Revolution* (Sacramento: State of California, 2010).

4. Based on materials originally developed by Charles P. Hammer and Lori D. Mann, *Britain Solves a Problem and Creates the Industrial Revolution* (Sacramento: State of California, 2010).

5. California Department of Education, *History/Social Science Content Standards for California Public Schools* (Sacramento: California Department of Education, 1998), 43.

Chapter 7

1. National Governors Association Center for Best Practices and the Council of Chief State School Officers, Common Core State Standards Initiative Mission Statement, http://www.corestandards.org.

2. National Governors Association Center for Best Practices and the Council of Chief State School Officers, *Common Core State Standards for English Language Arts & Literacy in History/Social Studies, Science, and Technical Subjects* (Washington, DC: National Governors Association Center for Best Practices and the Council of Chief State School Officers, 2010), 4.

3. National Governors Association Center for Best Practices and the Council of Chief State School Officers, *Common Core State Standards for Mathematics* (Washington, DC: National Governors Association Center for Best Practices and the Council of Chief State School Officers, 2010), 7.

4. Achieve, Inc., *Next Generation Science Standards* (Washington, DC: Achieve, Inc., 2013), 8.

5. Ibid., 93.

6. A complete catalog of EEI curriculum units in the standards they cover is available at: www.seer.org/EEI_Sources/Catalog.pdf.

7. Based on materials originally developed by Erica Peters and Dennis Silverio, "Water Works" (unpublished, Jackson Elementary School, Altadena, California, 2008).

8. These represent Pennsylvania academic standards in place at the time the instruction was planned and implemented.

9. California Department of Education, *Science Content Standards for California Public Schools: Kindergarten Through Grade Twelve* (Sacramento: California Department of Education, 1998), 25.

Chapter 8

1. Based on materials originally developed by Pamela Kattner and Hilary Heffner, "Watersheds" (unpublished, Seven Generations Charter School, Emmaus, PA); Council for Environmental Education, *WET in the City Curriculum and Activity Guide* (Houston: Council for Environmental Education, 2002), http://www.wetcity.org/resources.htm#WIC_Guide.

2. Gerald A. Lieberman et al., eds., "California's EEI Curriculum—Multiple Unit Titles," State of California, http://californiaeei.org/howTeachEEI/accessCurriculum.html.

3. Based on materials originally developed by Charles P. Hammer and Lori D. Mann, *Britain Solves a Problem and Creates the Industrial Revolution* (Sacramento: State of California, 2010).

4. Deborah Foss, *California's Natural Regions* (Sacramento: State of California, 2010).

5. Rachel Carson, *Silent Spring* (Boston: Houghton Mifflin, 1962); Albert Gore Jr., *An Inconvenient Truth: The Planetary Emergence of Global Warming* (Emmaus, PA: Rodale Press, 2006).

6. California Regional Environmental Education Community, CREEC Network, California Department of Education, http://www.creec.org.

7. The Acorn Group and State Education and Environment Roundtable, *Education and the Environment Initiative: Educator Needs Assessment* (Sacramento: State of California, 2005), www.seer.org/EEI_Sources/EEI_ENA.pdf.

Chapter 9

1. Based on materials originally developed by Charles P. Hammer and Lori D. Mann, *Britain Solves a Problem and Creates the Industrial Revolution* (Sacramento: State of California, 2010).

2. Based on materials originally developed by Erica Peters and Dennis Silverio, "Water Works" (unpublished, Jackson Elementary School, Altadena, CA, 2008).

3. Shirley M. Hord et al., *Taking Charge of Change* (Alexandria, VA: Association for Supervision and Curriculum Development, 1987).

4. Ibid.

5. Gerald A. Lieberman et al., *Implementing and Strengthening an EIC Model Program in Your School: A Self-Evaluation Guide* (Poway, CA: State Education and Environment Roundtable, 2000), 31.

Appendix A

1. California Environmental Protection Agency and California Integrated Waste Management Board, "California's Environmental Principles and Concepts," California Environmental Protection Agency and California Integrated Waste Management Board, http://www.calepa.ca.gov/education/principles/EPC.pdf.

Appendix B

1. Gerald A. Lieberman and Linda L. Hoody, *Closing the Achievement Gap: Using the Environment as an Integrating Context for Learning* (San Diego: State Education and Environment Roundtable, 1998).

2. Gerald A. Lieberman et al., *California Student Assessment Project, Phase One: The Effects of Environment-Based Education on Student Achievement* (San Diego: State Education and Environment Roundtable, 2000).

3. Gerald A. Lieberman et al., *California Student Assessment Project, Phase 2: The Effects of Environment-Based Education on Student Achievement* (San Diego: State Education and Environment Roundtable, 2005).

4. O. Bartosh, "Environmental Education: Improving Student Achievement" (master's thesis, The Evergreen State College, 2003), 79, http://www.seer.org/pages/research/Bartosh2003.pdf.

5. Edward Falco, South Carolina Department of Education, e-mail message to author with unpublished report "South Carolina 2004–05 PACT Scores, EIC vs. non-EIC Students," March 14, 2006.

6. Jeanette R. Wilson and Martha C. Monroe, "Biodiversity Curriculum That Supports Education Reform," *Applied Environmental Education and Communication* 4, no. 2 (2005): 125–138.

7. Emeka Emekauwa, *They Remember What They Touch . . . The Impact of Place-Based Learning in East Feliciana Parish* (Washington, DC: The Rural School and Community Trust, 2004), 5.

8. Kathy Abrams, "Summary of Project Outcomes from EE & SSS Schools' Final Report Data" (unpublished report, Florida Office of Environmental Education, Tallahassee, FL, 1999).

9. Ibid.

10. Ibid.

11. Ibid.

12. Ibid.

13. J. Glenn, *Environment-based Education: Creating High Performance Schools and Students* (Washington, DC: National Environmental Education and Training Foundation, 2000), http://www.neefusa.org/pdf/NEETF8400.pdf.

14. Ibid.

15. Ibid.

16. Ibid.

17. Ibid.

18. Ibid.

19. Ibid.

20. Ibid.

21. Allan Sterbinsky, *Rocky Mountain School of Expeditionary Learning Evaluation Report* (Memphis: Center for Research in Educational Policy, University of Memphis, 2002).

22. Ibid.

23. Tahoma High School, "Endeavor: An Integrated Learning Model for High School Students, 2008," Tahoma High School, Maple Valley, Washington, www.leadered.com/msc08/PowerPointsHandouts/Tahoma MSC PPT.ppt.

24. Ibid.

25. T. Taylor, "The Impact of Using the Environment as an Integrating Context for Improving Student Learning for Fourth-Grade Elementary School Students" (PhD dissertation, Argosy University, 2008), 1–29.

26. Lieberman and Hoody, *Closing the Achievement Gap.*

27. Lieberman et al., *California Student Assessment Project, Phase 1.*

28. Lieberman et al., *California Student Assessment Project, Phase 2.*

29. Bartosh, "Environmental Education," 79.

30. Falco, "South Carolina 2004–05 PACT scores."

31. Emeka Emekauwa, *The Star with My Name: The Alaska Rural Systemic Initiative and the Impact of Place-Based Education on Native Student Achievement* (Washington, DC: The Rural School and Community Trust, 2004).

32. Ibid.

33. Ibid.

34. Emekauwa, *They Remember What They Touch.*

35. Abrams, "Summary of Project Outcomes."

36. Glenn, *Environment-based Education.*

37. Ibid.

38. Ibid.

39. Ibid.

40. Michael Duffin et al., "Place-Based Education and Academic Achievement," PEER Associates, 2005, http://www.peecworks.org/PEEC/PEEC_Research/S0032637E.

41. Tahoma High School, "Endeavor."

42. Taylor, "The Impact of Using the Environment as an Integrating Context."

43. Lieberman and Hoody, *Closing the Achievement Gap.*

44. Lieberman et al., *California Student Assessment Project, Phase 1.*

45. Falco, "South Carolina 2004–05 PACT scores."

46. Emekauwa, *They Remember What They Touch.*

47. Glenn, *Environment-based Education.*

48. Duffin et al., "Place-Based Education and Academic Achievement."

49. Sterbinsky, *Rocky Mountain School of Expeditionary Learning Evaluation Report.*

50. Tahoma High School, "Endeavor."

51. Taylor, "The Impact of Using the Environment as an Integrating Context."
52. Lieberman and Hoody, *Closing the Achievement Gap.*
53. Lieberman et al., *California Student Assessment Project, Phase 1.*
54. Falco, "South Carolina 2004–05 PACT scores."
55. Emekauwa, *They Remember What They Touch.*
56. Glenn, *Environment-based Education.*
57. Taylor, "The Impact of Using the Environment as an Integrating Context."
58. Lieberman and Hoody, *Closing the Achievement Gap.*
59. Ibid.
60. Ibid.
61. Ibid.
62. Lieberman et al., *California Student Assessment Project, Phase 1.*
63. Edward Falco, *Environment-Based Education: Improving Attitudes and Academics for Adolescents* (Columbia: South Carolina Department of Education, 2004).
64. Julie Athman and Martha Monroe, "The Effects of Environment-Based Education on Students' Achievement Motivation," *Journal of Interpretation Research* 9, no. 1 (2004): 9–25.
65. Abrams, "Summary of Project Outcomes."
66. Ibid.
67. Ibid.
68. Lieberman and Hoody, *Closing the Achievement Gap.*
69. Ibid.
70. Ibid.
71. Julie (Athman) Ernst and Martha Monroe, "The Effects of Environment-Based Education on Students' Critical Thinking Skills and Disposition Toward Critical Thinking," *Environmental Education Research* 10, no. 4 (2004): 507–522.
72. Ibid.
73. Ibid.

About the Author

Dr. Gerald A. Lieberman is an internationally recognized authority on school improvement using natural and community surroundings as interdisciplinary contexts. Over the past thirty years, Dr. Lieberman has created and conducted professional development programs for more than nine thousand educators and other professionals, working with formal education systems at local, state, national, and international levels. He has designed and coordinated curriculum development programs in the United States, Costa Rica, Honduras, Colombia, and Argentina.

In 1995, Dr. Lieberman founded and has since directed the State Education and Environment Roundtable (SEER), initially a cooperative endeavor of departments of education in sixteen states sponsored by the Pew Charitable Trusts for its first ten years and administered by the Council of Chief State School Officers. In 1997, he led the development of the innovative educational strategy called the EIC Model, using a school's local Environment as the Integrating Context for learning. During the past fifteen years, Dr. Lieberman has focused his work on professional development programs that help schools achieve school improvement goals through implementation of the EIC Model.

From 2003 to 2010, Dr. Lieberman served as the principal consultant for the State of California's Education and the Environment Initiative (EEI), a cooperative endeavor of the California Environmental Protection Agency, the California Department of Education, the State Board of Education, the governor's secretary of education, the Integrated Waste Management Board, and the California Natural Resources Agency. Dr. Lieberman led the development of the state's "Environmental Principles and Concepts," the plan for the EEI curriculum approved by

California's State Board of Education in 2010 and now in use by K–12 classrooms throughout California.

Dr. Lieberman is the principal author of *Closing the Achievement Gap: Using the Environment as an Integrating Context for Learning*, a groundbreaking national study that received an award from the National Environmental Education Foundation for "bringing environmental learning into the mainstream of American K–12 education." He has authored forty other major reports, books, and articles on subjects related to environment-based teaching strategies and environmental science.

Dr. Lieberman received his PhD and MA from Princeton University and his BA from UCLA. He served on the executive committee of the National Education and Environment Partnership, operated by the National Environmental Education Foundation, and is a past chair of the Commission on Education and Communication of IUCN, the International Union for Conservation of Nature. He lives in Poway, California.

Index